S0-CDN-100

THE METHOD OF PAIRED COMPARISONS

The Method of Paired Comparisons

H. A. DAVID
Iowa State University

Monograph No. 41
Series Editor:
ALAN STUART

Second Edition, Revised

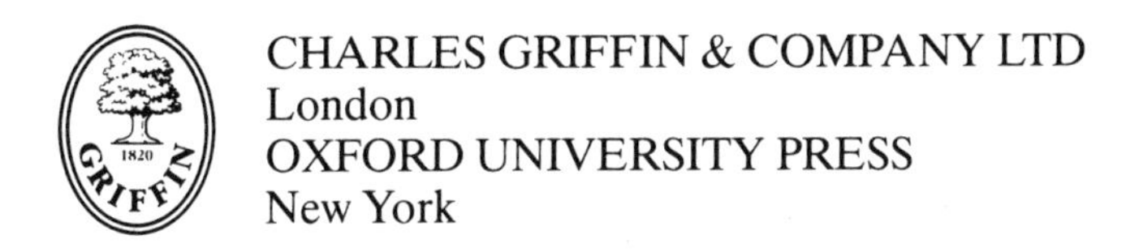

CHARLES GRIFFIN & COMPANY LTD
London
OXFORD UNIVERSITY PRESS
New York

CHARLES GRIFFIN & COMPANY LIMITED
16 Pembridge Road, London W11 3HL, Great Britain

First published 1963
Second edition 1988

Published in the U.S.A. by Oxford University Press,
200 Madison Avenue, New York, NY 10016

ISBN 0-19-520616-9

British Library Cataloguing in Publication Data

David, H. A.
The method of paired comparisons. —
2nd ed., rev. — (Griffin's statistical
monographs, no. 41).
1. Mathematical statistics
I. Title
519.5 QA276

ISBN 0 85264 290 3

Typeset in Great Britain by
Unicus Graphics Ltd, Horsham, West Sussex

Printed and bound in Great Britain by
Redwood Burn Limited, Trowbridge, Wilts

CONTENTS

PREFACE TO SECOND EDITION

Since the publication 25 years ago of the first edition of this monograph, the method of paired comparisons has seen much growth both in methodology and in applications. Fields of application now include acoustics, animal ecology, choice behaviour, dentistry, economics, epidemiology, food science, optics, preference testing, psychometrics, sensory testing, sports, and others.

In this edition we confine ourselves largely to methodological issues, using applications only for illustration. The new methods described are due primarily to statisticians, psychometricians, and mathematicians. Parametric approaches, especially those based on the Bradley–Terry model, have been greatly developed, allowing for much increased flexibility as well as multivariate generalizations. Techniques of mathematical programming have permitted analyses requiring milder model assumptions. Our distinctive emphasis continues to be on simple nonparametric procedures and on the analogy between paired comparisons and tournaments.

I am much indebted to Peter Bickel for suggesting the preparation of a second edition. The longstanding support of the U.S. Army Research Office is also gratefully acknowledged.

H. A. DAVID

1988

PREFACE TO FIRST EDITION

The method of paired comparisons has interested scientists with a variety of backgrounds. Important contributions to its development have been made not only by statisticians but also by psychologists—who were indeed the first to study it—by mathematicians, and by economists. Applications of the method have interested yet a wider group.

In this monograph I have attempted to give a unified picture of the present state of the subject. Apart from questions of design and analysis, the choice of appropriate probability models is discussed, and applications are indicated. The emphasis has been placed to some extent on simple, widely applicable procedures resting on a minimum of special assumptions. Writing as a statistician, I have avoided technical terms from other fields. The reader will find no mention of psychophysical continua or commodity bundles, but if these are his friends he will recognize them even though they are travelling incognito. A few footnotes giving textbook references have been inserted in the hope of helping the non-statistical reader over possible obstacles.

I am very grateful to Dr M. G. Kendall for encouraging me to undertake this account. It is a pleasure to acknowledge the help of Dr D. F. Morrison for his careful reading of the manuscript. For permission to reproduce tables or figures from their papers, I am indebted to R. C. Bose, R. A. Bradley, W. H. Clatworthy, L. R. Ford, Jr, S. M. Johnson, T. H. Starks, M. E. Terry, B. J. Trawinski, S. Ura, and the editors of the respective journals. The work was supported by the Army Research Office, Durham, North Carolina.

H. A. DAVID

1963

1 PROBABILITY MODELS

"Shall I compare thee to a summer's day?"

1.1 Introduction

In the method of paired comparisons objects are presented in pairs to one or more judges. We use "object" as a covering term that may stand for "person", "treatment", "stimulus", and the like. The basic experimental unit is the comparison of two objects, A and B, by a single judge who, in the simplest situation, must choose one of them. We shall say that the judge prefers this object although the choice will not necessarily represent a preference. More generally, the judge may be allowed to declare a tie, or asked to record a preference on some finer scale.

The comparison of A and B may be made by all the judges. If more than two objects are under consideration it is still easy to arrange that every judge performs every possible paired comparison. This situation may be called a "balanced paired-comparison experiment" and corresponds in the language of sport to a Round Robin tournament, the roles of the players in the tournament being analogous to those of the objects in the paired-comparison experiment. For t objects and n judges the number of paired comparisons will be $n\binom{t}{2}$.

The method of paired comparisons is used primarily in cases when the objects to be compared can be judged only subjectively; that is to say, when it is impossible or impracticable to make relevant measurements in order to decide which of two objects is preferable. As may be inferred, paired comparisons are widely employed by psychometricians. The method was indeed introduced in embryonic form by Fechner (1860, 1965) and, after considerable extensions, made popular by Thurstone (1927). Most frequent applications have been to sensory testing, especially taste testing, to consumer tests, personnel rating, and quite generally to the study of preference and choice behaviour. Of course, there are other methods of studying preferences, but we shall not enter into a detailed discussion of the individual merits of these methods, particularly as several extensive reviews are available (Bock and Jones, 1968; Coombs, 1964; Kendall, 1962; Torgerson, 1958). The method of paired comparisons is sometimes the obvious experimental procedure, as in testing various brands of razors where two kinds can be compared on a man's two cheeks. Again, in taste testing it is often difficult for a judge to cope with more than two tastes, and the introduction of a third may be thoroughly confusing.

In other instances the judge may be able to compare several objects at the same time. If this can easily be done a simple ranking of all objects may well be preferable. However, when differences between objects are small it is desirable to make the comparison between two of them as free as possible from any extraneous influences caused by the presence of other objects. Thus the method of paired comparisons has some advantages when a fine judgment is needed. It should be remembered that ranking is quick only when differences are fairly apparent; otherwise, the process of ranking requires in practice many repeated pairwise comparisons of tentative neighbours before a reasonable ordering becomes established. In these circumstances ranking becomes impracticable if there are many objects. Nor is it necessarily possible to achieve a wholly satisfactory ranking, a point which we shall elaborate shortly.

The allotment of grades or scores to the objects is another approach that comes to mind. If several grades are distinguishable, with a reasonable consensus among judges, such rating methods have the advantage that the scores can be treated roughly like measurements. The results can then often be interpreted without serious error by standard statistical techniques. Rating is liable to be much less successful when differences are small and when grades are hard to define. It is also possible to score a paired comparison on, say, a 7-point scale ($s = 3, 2, 1, 0, -1, -2, -3$) reading "strong preference for A", "preference for A", "slight preference for A", "no preference", "slight preference for B", etc. This is a compromise which retains the desirable close comparison between pairs of objects. Whereas in a rating procedure the grades are given to each object individually, making it hard to maintain consistency, the score s is a comparative score. Of course, its successful use depends on the existence of differences clear enough to be classified more finely than by a mere preference. The simplest form of paired comparison reduces the area of possible disagreement between judges to a minimum; even ties introduce difficulties, since some judges may declare ties more readily than others.

1.2 Pairwise comparisons of three objects

We consider first the case where three objects A, B, C are to be judged in pairs, since this simple situation brings out many of the essential features of paired-comparison experiments. Unless the contrary is stated, it will be assumed that each judgment consists of a simple preference for one or the other object. In particular, ties are not permitted.

Each of the three comparisons (AB), (AC), (BC) has two possible outcomes, so that there are eight distinguishable experiment results. Six of these are of the type

$$A \rightarrow B, A \rightarrow C, B \rightarrow C,$$

where the arrow $\rightarrow$ means "is preferred to", and may conveniently be represented by the partition(*) [210] of 3 which indicates that one object (not necessarily A) has scored two "wins", another one win, and the third none. The two remaining results

$$A \rightarrow B, B \rightarrow C, C \rightarrow A,$$

$$A \leftarrow B, B \leftarrow C, C \leftarrow A,$$

are represented by the partition [111] or $[1^3]$ and have been called *circular triads* by Kendall and Babington Smith (1940).

Clearly, a circular triad denotes an inconsistency on the part of the judge, and its simplest explanation is that the judge is at least partially guessing when declaring preferences. The judge may be guessing because of incompetence or because the objects are in fact very similar, the probability of a circular triad being $\frac{1}{4}$ when the objects are identical. But guessing is not the only explanation, for there may be no valid ordering of the three objects even when they differ markedly. Their merit may depend on more than one characteristic, and it is then somewhat artificial to attempt an

(*) We depart slightly from standard notation in showing zeros explicitly.

ordering on a linear scale. Under these circumstances the judge must mentally construct some function of the relevant characteristics and use this as a basis of comparison. It is not surprising that in complicated preference studies the function is vague and may change from one paired comparison to the next, especially when different pairs of objects may cause the judge to focus attention on different features of the objects. This last point helps to account for situations where a particular circular triad occurs frequently in repetitions of the experiment. However, circularity can occur even with a well-defined preference criterion based on two or more underlying dimensions (Tversky, 1969). An analogous occurrence is familiar in tournaments where, in repeated encounters between players or teams A, B, C, it may well happen that generally $A \rightarrow B$, $B \rightarrow C$, and yet C defeats A more often than not. The game of stone, scissors, and paper provides an extreme example.

It is a valuable feature of the method of paired comparisons that it allows such contradictions to show themselves, and this underlies many of the tests described in the next two chapters. Every practical precaution must be taken to ensure that the individual comparisons are independent or nearly so. This is not too difficult in taste testing, where the identity of the items to be tasted can be masked, and is aided by the fact that there will commonly be more than three items in the experiment. In personnel rating, on the other hand, there is a real danger that, due to a good memory, the judge's paired comparisons will degenerate into a ranking unless the number of people to be compared is very large. One partial way out of this dilemma is for a single judge to make only a fraction of all possible comparisons. This poses interesting problems of design (see Chapter 5). If, in a particular experiment, approximate independence has not been achieved, then the situation is intermediate between a straight ranking and independent paired comparisons. The analysis should therefore be made according to both methods. Only when the results agree can a conclusion be drawn with any comfort.

1.3 Basic models

The foregoing discussion may be formalized in a number of possible models which impose increasingly severe restrictions on the preference probabilities. Suppose that there are t objects A_1, $A_2, \ldots, A_t$ to be compared in pairs by each of n judges, the γth judge making r_γ replications of all possible $\binom{t}{2}$ comparisons. Let

$$x_{ij\gamma\delta}, \qquad i, j = 1, 2, \ldots, t, i \neq j; \quad \gamma = 1, 2, \ldots, n; \quad \delta = 1, 2, \ldots, r_\gamma;$$

be an indicator random variable taking the value 1 or 0 according as judge γ prefers A_i or A_j when making the δth comparison of the two. We assume throughout that all comparisons are statistically independent so that the $x_{ij\gamma\delta}$ are mutually independent, except that $x_{ij\gamma\delta} + x_{ji\gamma\delta} = 1$. Let

$$\text{(1.3.1)} \qquad \Pr(x_{ij\gamma\delta} = 1) = \pi_{ij\gamma\delta}.$$

The only restrictions on the π's are

$$\text{(1.3.2)} \qquad 0 \leqslant \pi_{ij\gamma\delta} \leqslant 1 \quad \text{and} \quad \pi_{ji\gamma\delta} = 1 - \pi_{ij\gamma\delta}.$$

Obvious special cases of interest are

$$\text{(1.3.3)} \qquad \pi_{ij\gamma\delta} = \pi_{ij\gamma} \quad \text{(no replication effect), and}$$

$$\text{(1.3.4)} \qquad \pi_{ij\gamma\delta} = \pi_{ij} \quad \text{(no replication or judge effect).}$$

If, as is often the case, replication effects are negligible but differences between judges are not, the last model is still very general as a description of individual preference ($\gamma = 1$). Possible effects due to the order of presentation of the objects in a pair are ignored at this stage (but see §7.4). When model (1.3.4) applies, the objects may be ranked according to the values of the average preference probabilities

$$\text{(1.3.5)} \qquad \pi_{i.} = \frac{1}{t-1} \sum_j{}' \pi_{ij},$$

the summation extending over all values of j other than $j = i$. Here $\pi_{i.} \geqslant \pi_{j.}$ does not imply $\pi_{ij} \geqslant \frac{1}{2}$; in fact, π_{ij} may take any value from 0 to 1. Another kind of ranking is possible if the following condition of *stochastic transitivity* holds for every triad of different objects A_i, A_j, A_k:

$$\text{(1.3.6)} \qquad \pi_{ij} \geqslant \tfrac{1}{2}, \quad \pi_{jk} \geqslant \tfrac{1}{2} \quad \text{imply} \quad \pi_{ik} \geqslant \tfrac{1}{2}.$$

Condition (1.3.6) leads to a clear-cut ordering of every triad and hence of all t objects. It is easy to see that this ranking does not necessarily agree with one made according to (1.3.5).

More stringent than (1.3.6) is the model of *strong stochastic transitivity*:

$$(1.3.7) \qquad \pi_{ij} \geqslant \tfrac{1}{2}, \quad \pi_{jk} \geqslant \tfrac{1}{2} \quad \text{imply} \quad \pi_{ik} \geqslant \max(\pi_{ij}, \pi_{jk}).$$

Intermediate to (1.3.6) and (1.3.7) is the condition of *moderate stochastic transitivity*:

$$(1.3.8) \qquad \pi_{ij} \geqslant \tfrac{1}{2}, \quad \pi_{jk} \geqslant \tfrac{1}{2} \quad \text{imply} \quad \pi_{ik} \geqslant \min(\pi_{ij}, \pi_{jk}).$$

For references on these three types of stochastic transitivity and for related conditions and their implications see Morrison (1963).

If to (1.3.7) we add:

$$(1.3.9) \qquad \pi_{ij} = \tfrac{1}{2} \quad \text{implies that } A_i \sim A_j \; (A_i \text{ is equivalent to } A_j);$$

then $\pi_{ik} = \pi_{jk}$ for $A_i \sim A_j$ and $\pi_{ik} = \pi_{ij}$ if $A_k \sim A_j$. Thus the objects A_i, A_j, A_k may be represented by points on a straight line, where the preference probability corresponding to the interval (A_k, A_i) is greater than that corresponding to the (proper) sub-

A_k A_j A_i

intervals (A_k, A_j), (A_j, A_i), and (1.3.9) takes care of coincident points. This representation can readily be extended to include all t objects on the one line. However, equal distances do not necessarily correspond to equal probabilities.

An interesting case where moderate but not strong stochastic transitivity may be expected to hold has been pointed out by Coombs (1959). To illustrate what is obviously a more general situation, let A_i, A_j, A_k be three grey objects of decreasing darkness. The judge is asked to express a preference in each paired comparison for the object considered the more *representative* grey. Here it seems reasonable to suppose that the judge has an ideal grey in mind and ranks other greys by their closeness to this ideal. If the relative locations on the scale of greyness are as shown below—

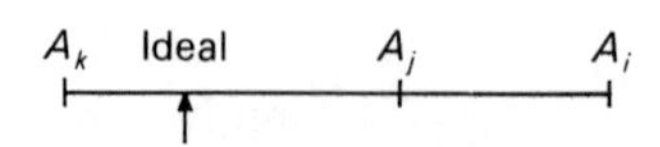

then $$\pi_{kj} > \tfrac{1}{2}, \quad \pi_{ji} > \tfrac{1}{2}, \quad \pi_{ki} > \pi_{kj},$$

but π_{ki} may well be smaller than π_{ji}, since comparisons on the same side of the ideal are less ambiguous than comparisons on opposite sides. This is confirmed experimentally by Coombs (1959). A similar effect may be anticipated when the three objects are £20, £25 and a watch of slightly greater (but unknown) value. See also Luce (1961) and Tversky (1969).

Linear model (for paired comparisons)

Suppose that the object A_i ($i=1, 2, \ldots, t$) has true value or "merit" V_i when judged on some characteristic. The merit of A_i, as registered by the judge, will vary from observation to observation and may be represented by the continuous r.v. y_i ($-\infty < y_i < \infty$). In a paired comparison of A_i and A_j the former will be preferred if $y_i > y_j$, the latter if $y_i < y_j$.

Let $z_i = y_i - V_i$, $i=1, 2, \ldots, n$. If the z_i are independent and identically distributed or, more generally, if every pair (z_i, z_j) has the same bivariate distribution, then $z_i - z_j$ must have the same distribution as $z_j - z_i$; i.e., $z_i - z_j$ is symmetrically distributed about zero. If we write $\Pr\{z_i - z_j < x\} = H(x)$, it follows that

$$\begin{aligned}\pi_{ij} &= \Pr\{y_i - y_j > 0\}\\ &= \Pr\{z_i - z_j > -(V_i - V_j)\}\\ &= H(V_i - V_j).\end{aligned}$$

Whenever the preference probabilities can be expressed in this manner, in terms of a symmetrical cdf, the y_i may be said to satisfy a *linear model.*

Since for H specified the π_{ij} depend only on the differences $V_i - V_j$, it follows that all π_{ij} are expressible as functions of $t-1$ independent differences of the V_i, with $\pi_{ij} \gtreqless \frac{1}{2}$ according as $V_i \gtreqless V_j$. The merits of the objects can therefore be usefully represented by t points on a linear scale with arbitrary origin, hence the term "linear". We may also say that the *characteristic* under study satisfies a linear model if the merit of *any* object (not only of the t objects A_i) can be represented on a linear scale.

The linear model is a generalization of the Thurstone–Mosteller model for which the y_i are assumed to be normal

$N(V_i, \sigma^2)$ variates, equi-correlated with common correlation coefficient ρ. This corresponds to taking

$$(1.3.10) \qquad \pi_{ij} = H(V_i - V_j) = \int_{-(V'_i - V'_j)}^{\infty} z(x)\,dx,$$

where $V'_i = \dfrac{V_i}{[2\sigma^2(1-\rho)]^{\frac{1}{2}}}$ and $z(x) = \dfrac{1}{\sqrt{(2\pi)}} e^{-\frac{1}{2}x^2}$.

Another important special case is provided by the Bradley–Terry model (Bradley, 1953) for which

$$(1.3.11) \qquad H(V_i - V_j) = \tfrac{1}{4} \int_{-(V_i - V_j)}^{\infty} \operatorname{sech}^2 \tfrac{1}{2}x\,dx,$$

where in Bradley's notation $V_i = \log \pi_i (\pi_i \geqslant 0, \Sigma \pi_i = 1)$; it follows that $\pi_{ij} = \pi_i/(\pi_i + \pi_j)$. Here the y_i may be taken to be independent doubly-exponential variates (Exercise 1.8).

Thompson and Singh (1967) arrive at the Thurstone–Mosteller model by regarding the *registered merit* or *experienced sensation* y as the average or median of a large number of signals transmitted to the judge's brain. Rather more artificially, they obtain the Bradley–Terry model when y is the maximum of a large number of transmitted signals having independent exponential-type distributions, so that y follows a doubly-exponential extreme-value distribution. Extensions of this approach to paired comparisons with ties and to triple comparisons are given by Beaver and Rao (1972). See also Singh and Gupta (1975).

Brunk (1960) has pointed out that for a linear model the merits ("worths" is his term) of the objects play a role somewhat analogous to "main effects" in the analysis of variance. To assume that they determine the preference probabilities completely is analogous to the hypothesis of no interaction in the analysis of variance. One may say that the V's represent the intrinsic merit of the objects in contrast to the opposite situation exemplified by (1.3.5), where the merit of an object is determined by the preference probabilities.

For any three objects A_i, A_j, and A_k subject to comparison under a linear model, it is possible to determine π_{ik} (say) from π_{ij}

and π_{jk}, provided only $H^{-1}(\pi_{ij})$ and $H^{-1}(\pi_{jk})$ are finite and uniquely defined. For,

$$V_i - V_k = (V_i - V_j) + (V_j - V_k) = H^{-1}(\pi_{ij}) + H^{-1}(\pi_{jk})$$

so that

$$\pi_{ik} = H[H^{-1}(\pi_{ij}) + H^{-1}(\pi_{jk})]. \tag{1.3.12}$$

In the case of the Bradley–Terry model, (1.3.12) can be simplified (see also Exercise 1.7 for a different approach). Latta (1979) calls a linear model based on H_1 *strictly more extreme* than one based on H_2 if, for any π_{ij}, π_{jk} in $(\frac{1}{2}, 1)$, π_{ik} is larger under H_1 than under H_2. He shows in particular that the Thurstone–Mosteller model is strictly more extreme than the Bradley–Terry model.

1.4 Extensions to multiple-choice situations

Suppose that not two but several objects are to be judged in a single comparison. The judge may, at one extreme, be asked merely to express a preference for one of the objects or, at the other extreme, to furnish a complete ranking. While not strictly within the scope of the method of paired comparisons, such multiple-choice situations provide an interesting extension. We mention here some attempts to construct probability models in special cases. In particular, the reader may refer to Luce (1959) who has studied in some detail a generalization of the Bradley–Terry model for paired comparisons and to Block and Marschak (1960), who have examined very thoroughly the logical interrelations of various paired and multiple comparison models. If M is *any* sub-group of the t objects which contains A_i, Luce postulates, in effect, that the probability of preferring A_i is

$$\pi_{i|M} = \frac{\pi_i}{\sum_M \pi_j}. \tag{1.4.1}$$

Clearly, this model, being much more sweeping than the linear Bradley–Terry model, can be expected to describe only rather special choice situations. In fact, (1.4.1) implies that if A_j is always preferred to A_i in a paired comparison of the two (so that $\pi_i = 0$) and if, however infrequently, A_i is sometimes preferred to A_k (so that $\pi_k = 0$), then A_k is never chosen from the set of three objects. Luce finds this consequence of linearity disturbing and proposes to remove it by restricting (1.4.1) to situations in which no $\pi_{ij} = 0$; if any $\pi_{ij} = 0$ he leaves $\pi_{i|M}$ undefined and postulates that

$$\pi_{j|M} = \frac{\pi_j}{\sum_{M'} \pi_i},$$

where M' is the set M with A_i removed. This seems a rather artificial attempt to extend the scope of the model since (1.4.1) still applies if $\pi_{ij} = \delta > 0$, for any δ, however small.

Model (1.4.1) (unextended) has the attractive if confining property of "independence from irrelevant alternatives", that is, if the sub-group M contains both A_i and A_j, then

$$\pi_{i|M}/\pi_{j|M} = \pi_i/\pi_j,$$

whatever other objects may be in M. As Debreu (1960) has pointed out, the model is inappropriate when, to take a simple case, M consists of two like objects A_1, A_2, and an unlike object A_3, corresponding to say $\pi_1 = \pi_2 = 0{\cdot}2$, $\pi_3 = 0{\cdot}3$. Here it seems more reasonable to take $\pi_{3|M} = 0{\cdot}3/0{\cdot}5$ rather than $0{\cdot}3/0{\cdot}7$, as required by (1.4.1). See also Exercise 1.11 and a review article by Luce (1977), that has a large number of references, especially to the psychological literature.

Triple comparisons

The comparison of objects in sets of three has received special attention. Pendergrass and Bradley (1960) have advocated and studied a model due to D. R. Cox according to which the objects A_i, A_j, A_k are ranked in that order with probability

$$\frac{\pi_i^2 \pi_j}{\pi_i^2(\pi_j + \pi_k) + \pi_j^2(\pi_k + \pi_i) + \pi_k^2(\pi_i + \pi_j)}. \tag{1.4.2}$$

Another model briefly considered by them gives this probability as

$$\frac{\pi_i \pi_j}{(\pi_i + \pi_j + \pi_k)(\pi_j + \pi_k)}, \tag{1.4.3}$$

leading to "very similar results in applications". For triple (and multiple) comparisons it would seem to be even more important than for paired comparisons to give the judge clear instructions as to how to do the rankings, if consistent results are to be obtained. If the judge first selects the treatment regarded as best of the three and then turns to picking the better of the remaining two, then model (1.4.3) is likely to be the more appropriate. Model (1.4.2) corresponds closely to the symmetrical situation where the judging is openly or tacitly by means of three paired comparisons which must be consistent. More than three such pairwise judgments may be required to achieve consistency, and the procedure is obviously less straightforward than the first one.(*) It is clear that if Luce's model holds, (1.4.2) and (1.4.3) are exact under the conditions stated, provided that all comparisons involved are independent. See also Beaver and Rao (1972).

1.5 Historical note on the method of paired comparisons

In his foreword to Torgerson's book *Theory and Methods of Scaling* Gulliksen speaks without ado of "Fechner's method of paired comparisons". Credit is rarely given to Fechner, yet there can be no doubt that the method goes back at least as far as the surprisingly little known account in his famous *Elemente der Psychophysik* published in 1860(†) and translated as *Elements of Psychophysics* in 1965(!). A few years prior to 1860 Fechner had carried out extensive experiments to examine Weber's law and especially to test the value of his own "Methode der richtigen und falschen Fälle" (method of right and wrong cases). In his main investigations he judged, many times and under a variety of circumstances, which of two vessels felt heavier, when there was in fact a small known difference D in mass (0·04 or 0·08 of the lighter vessel, whose mass P ranged from 300 to 3,000 g). The true result being known, as is typical of psychophysics in contrast to psychometrics, the number of right and wrong decisions could be tabulated. Initially, Fechner had an assistant to set up the vessels,

(*) Mallows (1957) used the same idea as the basis of a non-null model for the ranking of t objects.

(†) A symposium commemorating the centenary of the publication of this book appeared in the March 1961 issue of *Psychometrika*.

so that he himself was unaware of which was the heaver. He eventually abandoned this method and, peculiarly enough, in the experiments which he deemed worth reporting actually did all the arranging himself. Fechner's argument was that this allowed him to concentrate fully on the task of judgment before him and to obtain the most satisfactory results. His experimental technique is of some interest, although we must omit many details. To begin with, the lighter of the two vessels might be at the left. Fechner would pick it up in one second (with his left hand), set it down in another second, pause a second, and then raise and lower the other vessel (with his right hand), taking 5 seconds in all.[*] After 5 more seconds, time enough to record which vessel *felt* heavier, he repeated the procedure but began with the vessel to the right. In this fashion he completed one experimental unit of 32 comparisons, reversing the position of the vessels at the half-way stage. Fechner then did the same for different values of P and D, working steadily for about an hour. On other days the whole schedule was repeated until, for every pair of values of P and D studied, he had made a total of 2048 comparisons. Thes could be subdivided into 4 groups of 512 according to the position and order of lifting of the vessels ("space" and "time" effects).

Fechner was certainly not unaware of possible objections to his method of experimentation with foreknowledge but remarks sternly that he can accept such criticism only from those who have carefully tested the procedure for themselves.

In the analysis of his results Fechner's first step was to determine the variability of the apparent mass x of a vessel for any of the above series of experiments. He assumed x to be normally distributed about the true mass and equated the observed proportion of correct decisions to (in modern notation)

$$\int_{-\infty}^{D/2\sigma} \frac{1}{\surd(2\pi)} e^{-\frac{1}{2}u^2}\, du = \Phi(D/2\sigma), \tag{1.5.1}$$

allowing him to find $D/2\sigma$ from tables. With D known, this gave σ. It will be noted that if ρ is the correlation coefficient of the apparent masses of the two vessels, Mosteller's (1951a) approach gives the probability of a correct decision as $\Phi\{D/\sigma\surd[2(1-\rho)]\}$,

[*] In further experiments Fechner used left hand only and right hand only.

which includes (1.5.1) for the special case $\rho = -1$, not $\rho = 0$. This peculiarity arises from a rather unsatisfactory formulation of the problem (actually sanctioned by Möbius). Fortunately, it merely means that when determining σ, Fechner had in fact determined $\sigma/\surd[2(1-\rho)]$, i.e., $\sigma/\sqrt{2}$ for $\rho = 0$.

While the effects of space and time are ignored in the above, Fechner went further. Translating these effects into equivalent masses, he associated differences in mass of $D+S+T$, $D-S+T$, $D+S-T$, $D-S-T$ with four combinations of space and time. If the corresponding observed proportions of correct decisions were r_i/n $(i=1, \ldots, 4)$, he found from tables the equivalent normal deviates u_i and hence obtained the equations

$$u_1 = h(D+S+T), \qquad u_2 = h(D-S+T),$$

$$u_3 = h(D+S-T), \qquad u_4 = h(D-S-T),$$

where $1/h = \sigma\surd[2(1-\rho)]$ (for the general situation). Fechner determined h as $(u_1+u_2+u_3+u_4)/4D$, and S from

$$hS = \tfrac{1}{4}(u_1 - u_2 + u_3 - u_4).$$

Thus S is equivalent to a difference in mass of

$$\frac{u_1 - u_2 + u_3 - u_4}{u_1 + u_2 + u_3 + u_4} D,$$

independently of the value of ρ.

A more general account of Fechner's contributions to psychophysics is given in Chapter 7 of Stigler (1986).

Notes

(1) A formal treatment of models for preference and choice behaviour is given by, e.g., Luce (1959), Block and Marschak (1960), Pfanzagl (1971), and Falmagne (1985).

(2) There are many routes to the ubiquitous Bradley–Terry model. Indeed it was originally proposed by the noted mathematician E. Zermelo (1929) in a paper dealing with the estimation of the strengths of chess players in an uncompleted Round Robin tournament. However, Zermelo's excellent paper was an isolated piece of work. His model was independently rediscovered by

Bradley and Terry (1952) who demonstrated its usefulness in sensory testing. For a related approach see Good (1955).

(3) In place of preference probabilities one may work with *expected scores*. The two are equivalent when each paired comparison results in a score of 1 or 0, but expected scores cover also ties and other methods of scoring. See Brunk (1960).

(4) See also Bradley (1976) for a general summary of model formulations.

(5) The example of moderate stochastic transitivity on page 6 involving an ideal point is basic to Coombs's (1964) unfolding technique. A key issue is how to place the objects on a line in relation to the ideal point on the basis of paired-comparison data. The objects may in fact have to be placed in a space of two or more dimensions. We must refer the reader to the psychometric literature. In a recent article De Soete, Carroll and DeSarbo (1986) treat a multidimensional unfolding model and give many references.

EXERCISES

1.1 For a strong stochastic transitive model

$$1 \leqslant \pi_{ij} + \pi_{jk} + \pi_{ki} \leqslant 2.$$

1.2 If $\pi_{ij} \geqslant \frac{1}{2}$, $\pi_{jk} > \frac{1}{2}$ but $\pi_{ik} < \frac{1}{2}$, then the probability of observing a circular triad when A_i, A_j, A_k are compared exceeds $\frac{1}{4}$.

[Hint: The probability of a circular triad is expressible as

$$\tfrac{1}{4} + q_i q_j + q_j q_k + q_k q_i,$$

where $q_i = \pi_{ij} - \frac{1}{2}$, $\quad q_j = \pi_{jk} - \frac{1}{2}$, $\quad q_k = \pi_{ki} - \frac{1}{2}$.]

(Block and Marschak, 1960)

1.3 For a stochastic transitive model the probability of a circular triad is at most $\frac{1}{2}$.

(Block and Marschak, 1960)

1.4 Condition (1.3.7) necessarily implies a ranking according to (1.3.5), but (1.3.6) does not.

1.5 Let y_i $(i = 1, 2, \ldots, t)$ be independent $N(V_i, \sigma_i^2)$ variates. Then the preference probabilities $\pi_{ij} = \Pr(y_i > y_j)$ do not satisfy a linear model unless $\sigma_i^2 = \sigma^2$ (all i); in fact, the condition of strong stochastic transitivity does not necessarily hold.

[This situation corresponds to Thurstone's (1927b) Cases I–IV, the assumption of independence not being crucial.]

1.6 Define independent random variables y_1, y_2, y_3 by

$$\Pr\left(y_1 = \frac{-5}{3}\right) = \tfrac{3}{8}, \qquad \Pr(y_1 = 1) = \tfrac{5}{8},$$

$$\Pr(y_2 = -1) = \tfrac{5}{8}, \qquad \Pr\left(y_2 = \frac{5}{3}\right) = \tfrac{3}{8},$$

$$\Pr\left(y_3 = \frac{-2}{3}\sqrt{30}\right) = \tfrac{1}{16}, \quad \Pr(y_3 = 0) = \tfrac{7}{8}, \quad \Pr\left(y_3 = \frac{2}{3}\sqrt{30}\right) = \tfrac{1}{16}.$$

Verify that

$$\mathscr{E}(y_1) = \mathscr{E}(y_2) = \mathscr{E}(y_3) = 0,$$

$$\text{var } y_1 = \text{var } y_2 = \text{var } y_3 = \tfrac{5}{3},$$

and prove that

$$\pi_{21} = \pi_{32} = \pi_{13} = \tfrac{39}{64}.$$

[See Steinhaus and Trybula (1959), who state that $b = (\sqrt{5} - 1)/2$ is the largest value of b for which $\pi_{21} = \pi_{32} = \pi_{13} = b$ is possible. This result is proved by Trybula (1961) and Usiskin (1964). The latter in fact shows that $\pi_{12} = \pi_{23} = \ldots = \pi_{t-1,t} = \pi_{t1} = b(t)$ is possible for values of $b(t)$ that increase from $b(3) = 0{\cdot}61803$ to $b(4) = \tfrac{2}{3}, \ldots,$ $b(10) = 0{\cdot}73205, \ldots, b(\infty) = \tfrac{3}{4}$. See also Chang-Li-Chien (1961).]

1.7 Show that for the Bradley–Terry model the probabilities of the circular triads $A_i \to A_j \to A_k \to A_i$ and $A_i \leftarrow A_j \leftarrow A_k \leftarrow A_i$ are equal. Hence prove that π_{ik} is deducible from π_{ij} and π_{jk} by means of

$$\pi_{ik} = \frac{\pi_{ij}\pi_{jk}}{\pi_{ij}\pi_{jk} + \pi_{kj}\pi_{ji}}.$$

1.8 Let the independent variates y_i $(i=1, 2, \ldots, n)$ have extreme-value distributions with cdfs

$$F(y_i)=\exp[-\mathrm{e}^{-(y_i-V_i)}] \qquad -\infty<y_i, V_i<\infty.$$

Show that $x_{ij}=y_i-y_j$ $(j\neq i)$ has the logistic distribution with cdf

$$\Pr\{x_{ij}\leqslant x\}=\frac{1}{1+\mathrm{e}^{-(x-V_i+V_j)}} \qquad -\infty<x<\infty.$$

(Gumbel, 1961; Davidson, 1969)

1.9 If $\pi_i>\frac{1}{2}(\pi_j+\pi_k)$, then the probability of selecting A_i as the best of A_i, A_j, A_k is greater under model (1.4.2) than under (1.4.3).

1.10 Suppose that the three comparisons of the objects A_i, A_j, A_k are independent with preference probabilities satisfying the Bradley–Terry model. If the outcome is a circular triad the ranking is made at random. Show that A_i will be preferred with probability

$$\frac{\pi_i^2(\pi_j+\pi_k)+\frac{2}{3}\pi_i\pi_j\pi_k}{(\pi_i+\pi_j)(\pi_i+\pi_k)(\pi_j+\pi_k)},$$

and that A_i, A_j, A_k will be ranked in that order with probability

$$\frac{\pi_i^2\pi_j+\frac{1}{6}\pi_i\pi_j\pi_k}{(\pi_i+\pi_j)(\pi_i+\pi_k)(\pi_j+\pi_k)}.$$

1.11 Suppose that nt objects under comparison consist of t different objects $A_1, A_2, \ldots, A_t$, each replicated n times. Let $F(y_i-V_i)$ be the cdf of the characteristic y_i associated with A_i $(i=1, 2, \ldots, t)$. Show that the probability $\pi_{i,n}$ of selecting an A_i from this set is

$$\pi_{i,n}=\int_{-\infty}^{\infty}\left[\prod_{\substack{j=1\\ j\neq i}}^{t} F^n(y+V_i-V_j)\right]\mathrm{d}F^n(y).$$

Hence show that $\pi_{i,n}$ is independent of n iff $F(y)$ is of the form

$$F(y)=\exp[-\mathrm{e}^{-b(y-a)}] \quad -\infty<y, a<\infty,\ b>0.$$

[Thus the probability of selecting an A_i is unaffected by replication only for the Luce model.]

(Yellott, 1977)

2 COMBINATORIAL METHODS

2.1 Scores and circular triads

Suppose that a judge has made all $\binom{t}{2}$ paired comparisons of t objects. Ties are not permitted, so that the results may be expressed in the two-way preference table of 1's and 0's familiar from its use in tournaments. For $t = 5$ a typical outcome might be as shown:

	A_1	A_2	A_3	A_4	A_5	Score
A_1	—	0	1	1	1	3
A_2	1	—	1	0	0	2
A_3	0	0	—	0	0	0
A_4	0	1	1	—	1	3
A_5	0	1	1	0	—	2

The principal diagonal is left free and the entries below it are strictly speaking redundant. We see that A_1 has been preferred to A_3, A_4, A_5 but not to A_2, giving a row-sum score or simply *score*(*) of 3. In general, the number of preferences scored by A_i $(i = 1, 2, \ldots, t)$ will be denoted by a_i. Clearly

$$\sum_{i=1}^{t} a_i = \tfrac{1}{2}t(t-1) \tag{2.1.1}$$

and $2^{\frac{1}{2}t(t-1)}$ different preference tables are possible.

An alternative geometric representation, due like most of this section to Kendall and Babington Smith (1940), is sometimes

(*) This use of the term "score" should cause no confusion with its different meaning in Chapter 1.

more illuminating. We shall re-express this in the language of graph theory. The above table can be represented as a graph with *nodes* $A_1, A_2, \ldots, A_5$ and all its connecting *arcs* (Fig. 2.1). Arrows indicate the directions of the preferences, so that a_i is the number of arcs leaving A_i. Such a directed graph, or *digraph*, is said to be *complete* since all pairs of nodes are joined, and *asymmetric* since each arc is unidirectional.

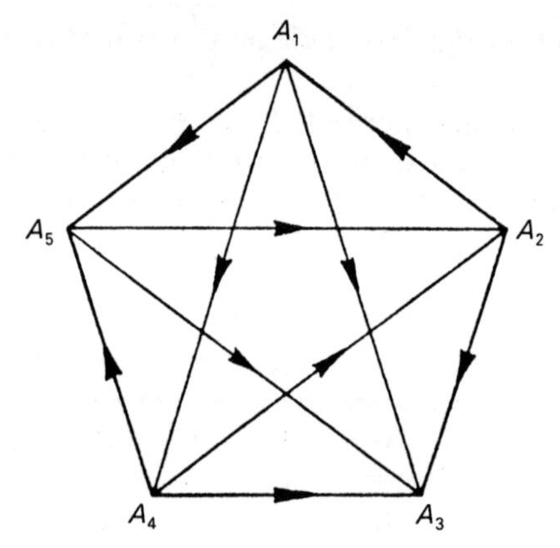

Fig. 2.1 A preference graph

Perhaps the first questions that come to mind on inspection of a preference table or the associated graph are:

(1) Has the judge been consistent in allocating preferences?
(2) Are there significant differences between the objects?

It is important to realize that these questions are closely related; for if there is no difference between the objects the judge cannot reasonably be expected to be consistent, while it is easy to be consistent if the differences are great. We shall refer to the null hypothesis that judgments are made at random ($\pi_{ij} = \frac{1}{2}$; all i, j; $i \neq j$), for whatever reason, as the "hypothesis of randomness".

For a group of three objects an inconsistency results in a circular triad; for a larger group one set of judgments may be regarded as more consistent than another if it includes fewer circular triads. Of the 10 triads in our example $A_1A_4A_2$ and $A_1A_5A_2$ are seen to be circular. It is clear that enumeration is inconvenient unless t is small. However, the number of circular triads c is deducible from the scores a_i by means of the relation

$$c = \frac{t}{24}(t^2 - 1) - \tfrac{1}{2}T, \tag{2.1.2}$$

where $T = \Sigma(a_i - \bar{a})^2$ and $\bar{a} = \Sigma a_i/t = \frac{1}{2}(t-1)$.

This result, first obtained by Kendall and Babington Smith (1940), can be easily proved by graph theory (Berge, 1962, p. 133). Note that a triad will not be circular if and only if two of its arcs issue from one of the triad's nodes, say A_i. Hence the total number of non-circular triads is

$$\begin{aligned}\sum_{i=1}^{t}\binom{a_i}{2} &= \tfrac{1}{2}(\Sigma a_i^2 - \Sigma a_i)\\ &= \tfrac{1}{2}T + \tfrac{1}{8}t(t-1)(t-3).\end{aligned}$$

Since the total number of triads is $\binom{t}{3}$, it follows that

$$\begin{aligned}c &= \tfrac{1}{6}t(t-1)(t-2) - \tfrac{1}{8}t(t-1)(t-3) - \tfrac{1}{2}T\\ &= \frac{t}{24}(t^2-1) - \tfrac{1}{2}T.\end{aligned}$$

Kendall and Babington Smith define the *coefficient of consistence* ζ by

$$\begin{aligned}\zeta &= 1 - \frac{24c}{t(t^2-1)}, \quad t \text{ odd},\\ &= 1 - \frac{24c}{t(t^2-4)}, \quad t \text{ even}.\end{aligned}$$

If and only if $\zeta = 1$ there are no inconsistencies in the configuration of preferences which is therefore immediately expressible as a ranking. As ζ decreases to zero the inconsistency as measured by the number of circular triads increases (Exercise 2.1).

Since T is a linear function of ζ the two are essentially equivalent. While T does not provide an immediate measure of consistency it appears to be more easily generalized to situations where each paired comparison is replicated n times. Thus if a_i now denotes the score of the ith object for the replicated experiment, we may simply define T_n as the total sum of squares of scores

$$T_n = \Sigma(a_i - \bar{a})^2, \tag{2.1.3}$$ (*)

so that $T = T_1$. Primarily for this reason we shall use the T-statistics in the sequel. Large values of T will lead to the rejection of the null hypothesis H_0 of randomness. In order to make tests of significance we shall show in §2.4 that T_n, when suitably standardized, is distributed on H_0 aproximately as a χ^2 with $t-1$ degrees of freedom.

Circular triads represent a very fundamental type of inconsistency. It does not follow that the number of circular triads or *3-cycles* provides a complete description of the circularity of a preference graph. Two such graphs may have the same number c of 3-cycles, and indeed the same score vector, and yet differ in the number of 4-cycles. However, we confine ourselves to c as a basic measure that is easy to determine. No simple formula is known for the number of 4-cycles.

Bauer (1978) has considered measuring circularity when not all $\frac{1}{2}t(t-1)$ paired comparisons have been made. There then arise triads $A_iA_jA_k$ in which A_i and A_j have not been compared and either $A_i \rightarrow A_k \rightarrow A_j$ or $A_i \leftarrow A_k \leftarrow A_j$. Bauer speaks of *reversals* in these cases and denotes the number of such triads by r. Clearly, on the assumption H_0 of randomly allocated preferences, each of these triads has probability $\frac{1}{2}$ of becoming a circular triad, were the comparison of A_i and A_j to be carried out. If each object is compared to $t-s$ other objects(†) $(s=1, 2, \ldots, t-1)$, Bauer obtains the formula

$$2c + r = K(s, t, D) - T, \tag{2.1.4}$$

where $K(s, t, D)$ is a constant that depends on s, t, and the design used, and $T = \Sigma(a_i - \bar{a})^2$ with $\bar{a} = \frac{1}{2}(t-s)$. For any (s, t, D), K can be

(*) This is, in fact, a special case of a statistic defined by Durbin (1951) for use in ranking experiments.

(†) Possible designs are discussed in Chapter 5.

found from the perfectly consistent case $c=0$, $r=0$. Since r is relatively easy to determine, c can be obtained from (2.1.4) but $c+\frac{1}{2}r$ would seem a better measure of circularity. For a fixed design, T can once again be used.

If there is a prior ordering of the objects with respect to the characteristic being judged (e.g., greyness in Coombs's example of §1.3), another type of inconsistency may occur. In the situation illustrated below, the results $A_i \to A_j$ and $A_k \to A_j$ are inconsistent

Increasing greyness →

A_i A_j A_k

(although they preclude the possibility of a circular triad); for $A_i \to A_j$ implies that the ideal I is to the left of A_j while $A_k \to A_j$ implies that I is to the right of A_j. Gerard and Shapiro (1958) have termed this type of inconsistency a *separation*. They show that the number of separations S in the pairwise comparison of t objects can be deduced from the preference table by means of the formula

$$S=\sum_{i=1}^{t} C_i(t-i-R_i), \tag{2.1.5}$$

where C_i, R_i are the ith column and row totals of the entries above the principal diagonal. Thus in the example at the beginning of this section $S=4$. The proof of (2.1.5) is immediate since C_i and $t-i-R_i$ are respectively the number of times objects to the left and right of A_i are preferred to A_i.

Coefficient of agreement

So far we have confined ourselves to the case of a single judge. If there are n judges each making $\binom{t}{2}$ comparisons the results can still be presented in a preference table, but now with entries α_{ij}, the number of times A_i was preferred to A_j. Then $\alpha_{ji}=n-\alpha_{ij}$. If the judges are in complete agreement, half of the α_{ij} equal n, the other half being zero. Note that the agreement may be complete even if there are inconsistencies present; conversely, lack of agreement does not imply the existence of inconsistencies.

Let

$$\Sigma = \sum_{i \neq j} \binom{\alpha_{ij}}{2},$$

the summation extending over $t(t-1)$ terms. Σ is the sum of the number of agreements between pairs of judges. Kendall and Babington Smith define the coefficient of agreement u by

$$u = \frac{2\Sigma}{\binom{n}{2}\binom{t}{2}} - 1. \tag{2.1.6}$$

If there is complete agreement, and only then, $u = 1$. The further we depart from this case, as measured by agreements between pairs of observers, the smaller u becomes. The minimum number of agreements occurs when each α_{ij} equals $\frac{1}{2}n$ if n is even or $\frac{1}{2}(n \pm 1)$ if n is odd. Correspondingly one has $u = -1/(n-1)$ or $-1/n$.

Σ may alternatively be expressed as

$$\Sigma = \tfrac{1}{2}\left[\sum_{i \neq j} \alpha_{ij}^2 - n\binom{t}{2}\right]$$

$$= \sum_{i<j} \alpha_{ij}^2 - n\sum_{i<j} \alpha_{ij} + \binom{n}{2}\binom{t}{2},$$

the last form being computationally convenient if the objects are arranged in order of decreasing scores, so that the α_{ij} will tend to be relatively small.

It is possible to test u for significance, the null hypothesis being that all the judges allot their preferences at random. This may be done (Kendall, 1962) from tables of the distribution of Σ for small combinations (t, n), and otherwise by treating

$$\frac{4}{n-2}\left[\Sigma - \tfrac{1}{2}\binom{t}{2}\binom{n}{2}\frac{n-3}{n-2}\right]$$

as an approximate χ^2 variate with degrees of freedom

$$\binom{t}{2}\frac{n(n-1)}{(n-2)^2}.$$

2.2 Other measures of inconsistency

We begin by showing that for a simple Round Robin tournament, c may be interpreted as the number of preference reversals necessary to break all ties in the score vector. Suppose $a_1 = a_2 = a$, with $A_1 \rightarrow A_2$. To see the effect of reversing this preference, let X be any other object. The only triads affected are those containing A_1 and A_2. There are four possible types of such triads:

$A_1 \rightarrow X \leftarrow A_2$, say x in number

$A_1 \leftarrow X \rightarrow A_2$,

$A_1 \rightarrow X \rightarrow A_2$, which must number $a-1-x$

$A_1 \leftarrow X \leftarrow A_2$, which must number $a-x$.

When $A_1 \rightarrow A_2$ is reversed, the first two remain noncircular. The third becomes circular, the fourth ceases to be so. Thus c has been reduced by 1. Continuing the tie-breaking process, we obtain the stated result, since $c=0$ when no ties are left. This proof, due to Kadane (1966), follows closely that used by Kendall and Babington Smith (1940) to establish (2.1.2).

After removal of all ties we arrive at a ranking of the objects, e.g., $(A_{i_1}, A_{i_2}, \ldots, A_{i_t})$, with A_{i_1} most preferred, where $(i_1, i_2, \ldots, i_t)$ is a permutation of $(1, 2, \ldots, t)$. Such a ranking, not necessarily unique, may be called a *nearest adjoining order.*

Slater's ***i***

More generally, one need not confine preference reversals to tied scores. Slater (1961) has proposed as a measure of inconsistency the minimum number $\boldsymbol{i}$ of preference reversals needed to reach a nearest adjoining order. Since c can be obtained in this manner also, but under restricted reversing, we have at once that $\boldsymbol{i} \leqslant c$ (Kadane, 1966). The probability distribution of $\boldsymbol{i}$ under randomness is given in Slater (1961) for $t \leqslant 8$.

The relative merits of c and $\boldsymbol{i}$ are illustrated in the following simple example. Suppose that in two sets of pairwise comparisons of the five objects $A_1, A_2, \ldots, A_5$ we have $A_i \rightarrow A_j$ for $i<j$, except that in the first set $A_3 \rightarrow A_1$ and in the second set $A_5 \rightarrow A_1$. In both cases there is a single inconsistency ($\boldsymbol{i}=1$) from the nearest adjoining order $(A_1, A_2, \ldots, A_5)$. On the other hand, the sets give rise to

the respective outcomes $[3^3 10]$, $[3^2 21^2]$ with different numbers of circular triads, namely $c = 1, 3$. Slater argues that every inconsistency should be given equal weight until H_0 (randomness) has been rejected. This is appropriate when inconsistencies are due entirely to lapses or accidental errors of judgment. Otherwise, for most reasonable alternatives H_a to randomness, the slight upset $A_3 \rightarrow A_1$ is more likely to occur than the major upset $A_5 \rightarrow A_1$. Therefore the first set, being more in accord with H_a than the second, is less compatible with H_0, as is reflected by its lower c-value.

It may be noted here that Ryser (1964) and Fulkerson (1965) have devised methods leading to rankings that minimize the number of inconsistencies subject to keeping row-sums monotone.

The case for Slater's ***i*** has been strengthened by a probabilistic basis provided in Thompson and Remage (1964) and Remage and Thompson (1966). If α_{ij} is the number of wins of A_i over A_j in n_{ij} comparisons, the likelihood function is

$$L = \prod_{i<j} \binom{n_{ij}}{\alpha_{ij}} \pi_{ij}^{\alpha_{ij}} (1 - \pi_{ij})^{n_{ij} - \alpha_{ij}} \qquad 1 \leqslant i < j \leqslant t. \tag{2.2.1}$$

Thompson and Remage propose ranking the objects by maximizing L with respect to the π_{ij}, subject to the weak stochastic transitivity condition that for any ordered triple $(A_{i_1}, A_{i_2}, A_{i_3})$

$$\pi_{i_1 i_2} \geqslant \tfrac{1}{2}, \quad \pi_{i_2 i_3} \geqslant \tfrac{1}{2}, \quad \pi_{i_1 i_3} \geqslant \tfrac{1}{2}.$$

They show that in the case $n_{ij} \leqslant 1 \, (1 \leqslant i < j \leqslant t)$, the resulting maximum-likelihood weak stochastic order is precisely Slater's nearest adjoining order.

Two other interpretations of ***i*** are given by Remage and Thompson (1966) in corresponding graph-theoretic language: The minimum number of arcs that must be removed, or alternatively the minimum number that must be reversed, in order to make an asymmetric digraph (not necessarily complete) free of cycles are both equal to ***i***.

Whereas the determination of c is straightforward for a simple Round Robin tournament in view of (2.1.2), the calculation of ***i*** (and a nearest adjoining order) continues to present a formidable challenge except when t is small. Interesting methods have been developed in Remage and Thomson (1966)(*) by dynamic programming, Phillips (1967, 1969) by ingenious elementary methods, and by Bezembinder (1981). Other authors have concentrated on obtaining the nearest adjoining orders under various generalizations. For unequal numbers of replications ($n_{ij} \geqslant 0$) deCani (1969) and Ranyard (1976) use linear programming and deCani (1972) a branch and bound algorithm. See also Hubert and Schultz (1975) and Hubert and Golledge (1981). Ties are incorporated as well in Singh and Thompson (1968), Flueck and Korsh (1974, 1975) and Singh (1976); cf. §7.3.

Bezembinder's ρ

A new index of circularity, ρ, intended to cover also incomplete paired-comparison data, was introduced in Bezembinder (1981) which should be consulted for a fuller treatment. We shall illustrate the construction of ρ in the very simple case $t=5$. The nine possible score vectors or partitions are shown in Table 2.1,

Table 2.1 Values of c and Bezembinder's (1981) ρ for the possible score vectors in a simple paired-comparison experiment involving 5 objects

Score vector	c	2-partition	ρ
[43210]	0	(11111)	0
431^3	1	311	1
42^30	1	311	1
42^21^2	2	41	3
3^310	1	311	1
3^22^20	2	41	3
3^221^2	3	5	6
32^31	4	5	6
2^5	5	5	6

together with the corresponding values of c obtained from (2.1.2) and needed here only for comparison. The preference table at the beginning of this chapter is included under $[3^22^20]$, the notation being more fully explained in §2.3.

(*) A computer program is given by Loflin and McMahan (1982).

In further description it is convenient to switch to the language of Round Robin tournaments. At the conclusion of such a tournament the players can always be arranged into disjoint sets T_1, $T_2, \ldots, T_k$ (for some k, $1 \leqslant k \leqslant t$) with the following properties (cf. Kadane, 1966):

(a) Each player in T_h has defeated all players in $T_{h'}$ for all $h < h'$ $(h, h' = 1, 2, \ldots, k)$;

(b) For any two players A_i, A_j in the same set T_h, either $A_i \rightarrow A_j$ or there exist other players $A_{i_1}, A_{i_2}, \ldots$, in T_h such that $A_i \rightarrow A_{i_1} \rightarrow A_{i_2} \rightarrow \ldots \rightarrow A_j$. Such a set is called a *strong* subtournament.

If we wish to rank the players on the basis of the tournament results only, then clearly players in T_h should rank ahead of those in $T_{h'}$ for $h < h'$. (We are not concerned here with questions of statistical significance.) In the extreme case $k = t$ we obtain a *complete ordering* of all the players. However, at the other extreme, $k = 1$, there is no obviously best way of ranking the players. For $1 < k < t$ the sets T_1, $T_2, \ldots, T_k$ provide a *partial ordering* in which only the rankings within sets remain in doubt.

The third column of Table 2.1 gives the unique partition of the players into strong subtournaments corresponding to the nine possible tournament score vectors. For example, $[3^3 10]$ represents the following hierarchy of players: 3 who defeated the same 2 players but tied among themselves, 1 who defeated only 1 player, and 1 who scores 0, resulting in the partition (311). Players in the same strong (sub-) tournament need not have the same score, e.g., $[3^2 21^2]$ and $[32^3 1]$, each of which represents a strong tournament.

The reason for the name $\bar{2}$-partition is that strong components of size 2 are obviously impossible (but any other size is possible since there are k-cycles for $k = 3, 4, \ldots$). There is a simple method for determining the $\bar{2}$-partitions in complete tournaments; e.g., for $[42^2 1^2]$ expressed as $a_{(1)} \leqslant a_{(2)} \leqslant \ldots \leqslant a_{(5)}$, we have

i	1	2	3	4	5
$\sum_{j=1}^{i} a_{(j)}$	1	2	4	6	10
$\binom{i}{2}$	0	1	3	6	10

When equality of $\sum_{j=1}^{i} a_{(j)}$ and $\binom{i}{2}$ is attained for the first time, for $i = r$ (say), it is clear that the players with the r lowest scores have not defeated any other players, thus forming a strong component of size r. Similarly, when equality is next attained, for $i = s$, a second component of size $s - r$ results, etc. (cf. Exercise 2.9). Thus $[42^2 1^2]$ gives the partition (41).

Next, let N_j be the number of pairs of nodes in the jth component, of size t_j ($j = 1, 2, \ldots, \nu$), that are joined by an arc. Then Bezembinder's measure is

$$(2.2.2) \qquad \rho = \nu - t + \sum_{j=1}^{\nu} N_j.$$

In the case of a complete tournament, $N_j = \frac{1}{2} t_j (t_j - 1)$. From (2.2.2) one obtains the last column of Table 2.1.

We see that the $\bar{2}$-partitions are informative about the nature of the tournament outcome but fail to distinguish between different outcomes within a $\bar{2}$-partition. This is, of course, reflected by ρ. To some extent c provides such distinctions and does so increasingly as t increases. It should be noted that while the number of $\bar{2}$ partitions grows with t, the probability $P(t)$, under the null hypothesis of randomness, that a tournament of t players is strong quickly approaches 1 (e.g., $P(10) = 0{\cdot}962$, $P(15) = 0{\cdot}998$; see Moon, 1967, p. 3). On the other hand, different $\bar{2}$-partitions may produce the same value of c.

2.3 Basic distribution theory

As we have seen, the results of a paired-comparison experiment can be summed up in the vector of scores $(a_1, a_2, \ldots, a_t)$, where $\Sigma a_i = \frac{1}{2} t(t-1)$. For many purposes the numbering of the objects is of no consequence and $(a_1, a_2, \ldots, a_t)$ may be replaced by the partition $[x_1^{r_1} x_2^{r_2} \ldots x_m^{r_m}]$, where the x's are simply a rearrangement of the a's in decreasing order of magnitude, and r_1 is the number of occurrences of the largest score x_1, etc. It follows that

$$\sum_{u=1}^{m} r_u = t, \qquad \sum_{u=1}^{m} r_u x_u = \tfrac{1}{2} t(t-1).$$

The probability distribution of the permissible partitions is of basic importance since from it can be deduced the distribution of all test-statistics, such as T, which are functions of the scores. When the hypothesis of randomness is true the frequencies of the various partitions give equivalent information since they can be converted into probabilities on division by $2^{\frac{1}{2}t(t-1)}$.

A possible approach to the enumeration problem involved is to consider the generating function

$$G(t) = \prod_{i<j} (b_i + b_j), \tag{2.3.1}$$

a product of $\frac{1}{2}t(t-1)$ terms which correspond to the comparisons made. The expansion of $G(t)$ contains $2^{\frac{1}{2}t(t-1)}$ terms, the exponents in each of which provide a possible set of scores. For example,

$$G(3) = (b_1 + b_2)(b_1 + b_3)(b_2 + b_3) = \Sigma b_1^2 b_2 + 2b_1 b_2 b_3, \tag{2.3.2}$$

the sum being over 6 terms. Thus there are 6 outcomes of type [210] and 2 of type $[1^3]$. $G(t)$ is symmetric in the b's and may be expanded as in (2.3.2) as a sum of monomial symmetric functions. In Table 2.2 partitions of scores and their frequencies are listed for $t \leqslant 6$. The table has been constructed and can be extended (David, 1959) by noting that

(a) $$G(t+1) = G(t) \prod_{i=1}^{t} (b_{t+1} + b_i)$$

$$= G(t)[b_{t+1}^t + (1)b_{t+1}^{t-1} + (1^2)b_{t+1}^{t-2} + \ldots + (1^t)],$$

where $(1^s) = \Sigma b_1 b_2 \ldots b_s, \quad (s = 1, 2, \ldots, t)$;

(b) the partitions

$$[x_1^{r_1} x_2^{r_2} \ldots x_m^{r_m}] \quad \text{and} \quad [(t-1-x_m)^{r_m} \ldots (t-1-x_2)^{r_2}(t-1-x_1)^{r_1}]$$

have the same frequencies since one is obtainable from the other by interchanging wins and losses.

Three generalizations of (2.3.1) are perhaps worth pointing out, although they do not necessarily provide an easy solution to the problems of enumeration involved.

Table 2.2 Partitions of scores and their frequencies in a simple paired-comparison experiment involving *t* objects

From David (1959) by permission of the Editor of *Biometrika*

$t=3$	
[210]	6
1^3	2
Total	8

$t=4$	
[3210]	24
31^3	8
2^30	8
2^21^2	24
Total	64

$t=5$	
[43210]	120
431^3	40
42^30	40
42^21^2	120
3^310	40
3^22^20	120
3^221^2	240
32^31	280
2^5	24
Total	1024

$t=6$	
[543210]	720
5431^3	240
542^30	240
542^21^2	720
53^310	240
53^22^20	720
53^221^2	1440
532^31	1680
52^5	144
4^3210	240
4^31^3	80
4^23^210	720
4^232^20	1440
4^2321^2	2880
4^22^31	1680
43^320	1680
43^31^2	1680
43^22^21	8640
432^4	2400
3^22^3	2640
3^421	2400
3^50	144
Total	32 768

(i) If for all $i \neq j$ not merely one but n_{ij} comparisons of A_i and A_j are made, the appropriate generating function is

$$\prod_{i<j} (b_i + b_j)^{n_{ij}}. \tag{2.3.3}$$

(ii) If $A_i \rightarrow A_j$ with probability π_{ij} for all $i \neq j$, then $G(t)$ can be replaced by

$$\prod_{i<j} (b_i \pi_{ij} + b_j \pi_{ji}) \tag{2.3.4}$$

to generate the *probabilities* of the various partitions.

(iii) Suppose that instead of being two-point (1 and 0) the scoring scale is $(r+1)$-point $\{1, (r-1)/r, \dots, 1/r, 0\}$, with the total score in any one comparison still unity. Then

$$\prod_{i<j} (b_i + b_i^{(r-1)/r} b_j^{1/r} + b_i^{(r-2)/r} b_j^{2/r} + \dots + b_j)$$

generates all possible outcomes. This includes the case where ties as well as clear preferences are permitted $(r=2)$.

The first generalization is of special interest for $n_{ij} = n$, all $i \neq j$, since this corresponds to a balanced paired-comparison experiment with n replications. The generating function $G_n(t)$ for this case is simply $[G(t)]^n$, and tables of partitions and their frequencies for parameters (t, n) may be built up from those for $(t, 1)$ in Table 2.2. Trawinski (1961) has tabulated all cases up to (3, 20), (4, 7), and (5, 3), in addition to $n=1$, $t \leqslant 8$. Closely related tables had been given earlier by Bradley and Terry (1952) and Bradley (1954b).

The approach of this section was anticipated by Rapoport (1949) and Landau (1951) in papers dealing with the theory of peck rights among hens.

2.4 Distribution of the scores and of related functions

In this section we shall obtain the null distributions of various functions of the scores in a balanced paired-comparison experiment of t objects and n replications. The results will be applied in Chapter 3 to the construction of a series of tests of significance for equality of the t objects.

Single score

Since under the hypothesis H_0 of randomness object A_i has probability $\frac{1}{2}$ of being preferred in each of its $n(t-1)$ comparisons with the other $t-1$ objects, the score a_i is a binomial $b(\frac{1}{2}, n(t-1))$ variate.

Joint distribution of scores

Consider first the case $n=1$. To find the joint distribution of two scores, which without loss of generality may be taken to be a_1

and a_2, note that two cases may be distinguished according as $A_1 \to A_2$ or $A_1 \leftarrow A_2$. If $A_1 \to A_2$ then A_1, A_2 must be preferred respectively to $a_1 - 1$, a_2 of the remaining $t-2$ objects, etc. Thus

$$(2.4.1) \qquad f(a_1, a_2) = \frac{1}{2^{2t-3}}\left[\binom{t-2}{a_1-1}\binom{t-2}{a_2} + \binom{t-2}{a_1}\binom{t-2}{a_2-1}\right].$$

The argument may be extended to give the joint distribution of a_1, $a_2, \ldots, a_s$ $(s<t)$ by a consideration of the $2^{\frac{1}{2}s(s-1)}$ outcomes of the paired comparisons of s objects only. If the scores in this sub-experiment are a_1', $a_2', \ldots, a_s'$, the final scores (when all t objects are involved) will be $a_1, a_2, \ldots, a_s$ with probability

$$(2.4.2) \quad f(a_1, a_2, \ldots, a_s \mid a_1', a_2', \ldots, a_s')$$

$$= \frac{1}{2^{s(t-s)}}\binom{t-s}{a_1-a_1'}\binom{t-s}{a_2-a_2'}\cdots\binom{t-s}{a_s-a_s'},$$

provided we use the convention

$$\binom{n}{r} = 0 \qquad \text{for} \qquad r<0 \quad \text{and} \quad r>n.$$

It follows that

$$(2.4.3) \qquad f(a_1, a_2, \ldots, a_s) = \frac{1}{2^{\frac{1}{2}s(2t-s-1)}}\Sigma\prod_{r=1}^{s}\binom{t-s}{a_r-a_r'},$$

where the sum is, in effect, over all outcomes of the sub-experiment compatible with the final scores. For $n \geqslant 1$, (2.4.3) generalizes to

$$(2.4.4) \qquad f(a_1, a_2, \ldots, a_s) = \frac{1}{2^{\frac{1}{2}ns(2t-s-1)}}\Sigma\prod_{r=1}^{s}\binom{n(t-s)}{a_r-a_r'}.$$

An alternative approach is given in §2.5.

Difference of two scores

The distribution $f(d)$ of $d=a_1-a_2$ can, of course, be obtained from the joint distribution of a_1 and a_2. $f(d)$ tends to normality with increasing n or t. This follows from the fact (Exercise 2.4) that the characteristic function $\phi(u)$ of $d/\sqrt{(\frac{1}{2}nt)}$ is

(2.4.5) $$\phi(u) = [\cos \tfrac{1}{2}u\sqrt{(2/nt)}]^{2n(t-2)}[\cos u\sqrt{(2/nt)}]^n,$$

which tends to $e^{-\frac{1}{2}u^2}$ as n or t becomes large.

We now find an exact expression for

$$P_{tnm} = \Pr(|a_1 - a_2| \geqslant m|H_0),$$

where m is a positive integer. We have

$P_{tnm} = 2\Pr(a_1 - a_2 \geqslant m|H_0)$

$= 2 \sum_{r=m}^{n(t-1)} \sum_{p=0}^{n}$ [Pr (in all comparisons between objects A_1 and A_2, A_1 is preferred $(2p-n)$ more times than is A_2) $\times$ Pr(A_1 is preferred $(r-2p+n)$ more times than is A_2 in comparisons with the other $(t-2)$ objects]

$$= 2 \sum_{r=m}^{n(t-1)} \sum_{p=0}^{n} \binom{n}{p} 2^{-n} \sum_{q=r-2p+n}^{n(t-2)} \binom{n(t-2)}{q} \times \binom{n(t-2)}{q-r+2p-n} 2^{-2n(t-2)}$$

$$= 2^{3n-2nt+1} \sum_{r=m}^{n(t-1)} \sum_{p=0}^{n} \binom{n}{p} \sum_{q=r-2p+n}^{n(t-2)} \binom{n(t-2)}{q} \binom{n(t-2)}{q-r+2p-n}.$$

(2.4.6)

The equality of the coefficients of x^{r-2p+n} in the expansion of the two sides of the identity

$$(1+x)^{n(t-2)}(1+x^{-1})^{n(t-2)} \equiv (1+x)^{2n(t-2)}\, x^{-n(t-2)}$$

gives

(2.4.7) $$\sum_{q=r-2p+n}^{n(t-2)} \binom{n(t-2)}{q} \binom{n(t-2)}{q-r+2p-n} = \binom{2n(t-2)}{n(t-3)-r+2p}.$$

Substituting (2.3.7) into (2.3.6), we obtain

$$P_{tnm} = 2^{3n-2nt+1} \sum_{r=m}^{n(t-1)} \sum_{p=0}^{n} \binom{n}{p} \binom{2n(t-2)}{n(t-3)-r+2p}.$$

Top score x_1

For values of t and n within the range of the basic tables of §2.3 it would be easy to build up the probability distribution of x_1. Usually this distribution is required only for large values of x_1. We therefore describe a method which is convenient in this case and also applicable for larger values of t and n.

Let E_i be the event $a_i \geqslant M (0 \leqslant M \leqslant n(t-1))$. On H_0 it is clear that

$$\Pr(E_{i_1} E_{i_2} \dots E_{i_s}) = \Pr(E_1 E_2 \dots E_s).$$

Then the principle of inclusion and exclusion yields (e.g., Feller, 1967, §4.1)

$$(2.4.8) \quad \Pr(x_1 \geqslant M) = Pr\left(\bigcup_{i=1}^{t} E_i\right) = \sum_{s=1}^{t} (-1)^{s-1} \binom{t}{s} \Pr(E_1 E_2 \dots E_s).$$

For large values of M, the later terms in the sum will be zero since only a few scores can be high. In particular, only the top score can exceed $n(t-1) - \frac{1}{2}n$, so that for $M > n(t-1) - \frac{1}{2}n$

$$(2.4.9) \quad \Pr(x_1 \geqslant M) = \binom{t}{1} \Pr(E_1) = t.2^{-n(t-1)} \sum_{k=M}^{n(t-1)} \binom{n(t-1)}{k}.$$

The exact evaluation of $\Pr(x_1 \geqslant M)$ is much more laborious for smaller values of M, but a simple approximation is often useful. Provided M is large enough, so that $\Pr(E_1) < 1/t$, one can apply Bonferroni's inequalities to obtain

$$t\Pr(E_1) - \binom{t}{2} \Pr(E_1 E_2) \leqslant \Pr(x_1 \geqslant M) \leqslant t\Pr(E_1).$$

Since the sum of the scores is a constant, we have

$$\Pr(E_1 \mid E_2) \leqslant \Pr(E_1),$$

and therefore,
$$\Pr(E_1 E_2) \leqslant [\Pr(E_1)]^2.$$

It follows that

$$(2.4.10) \quad t\Pr(E_1) - \binom{t}{2} [\Pr(E_1)]^2 \leqslant \Pr(x_1 \geqslant M) \leqslant t\Pr(E_1).$$

Since $\Pr(E_1)$ can be calculated directly or found in a table of the binomial distribution, this gives easily obtained bounds on $\Pr(x_1 \geqslant M)$. Thus if $t\Pr(E_1) = \beta$ one has

$$\beta - \frac{(t-1)}{2t}\beta^2 \leqslant \Pr(x_1 \geqslant M) \leqslant \beta,$$

and the range between the two bounds decreases as t or β decrease.

Sum of squares of scores

Since a_i is $b(\frac{1}{2}, n(t-1))$ on H_0 it follows that

$$d_i = 2[a_i - \tfrac{1}{2}n(t-1)]/\surd(nt) \tag{2.4.11}$$

has mean zero and variance $\sigma^2 = (t-1)/t$. Also, knowing that $\Sigma d_i = 0$ and that the d_i are by symmetry equi-correlated with correlation coefficient ρ (say), we have

$$\operatorname{var}(\Sigma d_i) = t\sigma^2 + 2\binom{t}{2}\rho\sigma^2 = 0,$$

so that $\rho\sigma^2 = -1/t$. The dispersion matrix Σ_d of the variable $\mathbf{d} = (d_1, d_2, \ldots, d_t)$ is therefore a $t \times t$ matrix with $(t-1)/t$ along the principal diagonal and $-1/t$ elsewhere. From the multivariate central limit theorem (see, e.g., Anderson, 1984) the asymptotic distribution of $\mathbf{d}$ as $n \to \infty$ is normal $N(\mathbf{0}, \Sigma_d)$. Now

$$\begin{vmatrix} \frac{t-1}{t} - \lambda & -\frac{1}{t} & & -\frac{1}{t} \\ -\frac{1}{t} & \frac{t-1}{t} - \lambda & & \\ \cdot\ \cdot\ \cdot & & & \cdot\ \cdot\ \cdot \\ -\frac{1}{t} & -\frac{1}{t} & & \frac{t-1}{t} - \lambda \end{vmatrix} = -\lambda(1-\lambda)^{t-1},$$

so that Σ_d has $(t-1)$ characteristic roots equal to one and the remaining root equal to zero. As is shown in Cramér (1946,

§24.5) the above results are sufficient to prove that

(2.4.12) $$D_n = \Sigma d_i^2 = 4T_n/(nt)$$

has a limiting χ^2-distribution with $(t-1)$ DF as $n \to \infty$.

D_n may indeed be expected to follow an approximate χ^2_{t-1} distribution, provided $n(t-1)$ is not too small; for the d_i are approximately normal $N(0, (t-1)/t)$ and Σd_i^2 may be transformed by an orthogonal transformation into $\sum_{j=1}^{t-1} y_j^2$, where the y_j are uncorrelated and approximately unit normal.

The χ^2-approximation to D_n, suggested by Durbin (1951), has been examined by Starks (1958) by comparison with exact results obtainable from the basic tables referred to in §2.3. It turns out to be reasonably satisfactory in the region of the upper 10 per cent point and beyond, and may confidently be used outside the range of Appendix Table 1 constructed from the basic tables (Trawinski, 1961). Surprisingly, this simple approximation is in general quite as effective near the upper percentage points as are more complicated approximations that have been suggested.

Range of scores

From the preceding we may expect the distribution of the range of the d_i to be asymptotically the same as that of the range W_t of t independent normal variates with variance

$$\sigma^2(1-\rho) = (t-1)/t + 1/t = 1$$

(cf. Hartley, 1950). The probability integral of W_t is given in Table 23 of *Biometrika Tables for Statisticians*, Vol. 1 (Pearson and Hartley, 1970). Hence $\Pr(W_t \geqslant 2R/\surd(nt))$ is an easily obtained approximation to Pr (range $a_i \geqslant R$) and will improve in accuracy as $n(t-1)$ increases. It has been found empirically that a better approximation results from applying a continuity correction (of half the usual size) giving $\Pr(W_t \geqslant (2R - \frac{1}{2})/\surd(nt))$ as the approximating function.

Ties

The null distribution of scores and related functions in the presence of ties has been examined by Gillot and Caussinus

(1966), the score a_i being now the number of wins plus half the number of ties of A_i. For the comparison of A_i and A_j it is assumed that

$$\Pr\{A_i \to A_j\} = \Pr\{A_j \to A_i\} = \pi \leqslant \tfrac{1}{2} \quad (i, j = 1, 2, \ldots, t;\ i \neq j)$$

The authors show, for example, that in generalizations of (2.4.12) the ratio

$$2(t-1)\Sigma(a_i - \bar{a})^2/N,$$

where N = total number of untied games, tends to a χ^2_{t-1}-distribution as $n \to \infty$.

2.5 Distribution theory in the non-null case

Primarily with a view to applications in Chapter 6, we now consider the non-null case, essentially model (1.3.4), for which

$$\Pr(x_{ij\gamma} = 1) = \pi_{ij} \quad (i, j = 1, 2, \ldots, t;\ i \neq j;\ \gamma = 1, 2, \ldots, n).$$

Joint distribution of the scores

Let α_{pr} $(p > r)$ be the number of times A_p is preferred to A_r in n comparisons. The joint distribution of these independent variates is given by

$$f(\alpha_{t,t-1}, \alpha_{t,t-2}, \ldots, \alpha_{21}) = \prod_{p>r}^{t} \binom{n}{\alpha_{pr}} \pi_{pr}^{\alpha_{pr}} \pi_{pr}^{n-\alpha_{pr}}. \tag{2.5.1}$$

Since the scores may be expressed as

$$\begin{aligned}
a_t &= \alpha_{t1} + \alpha_{t2} + \ldots && + \alpha_{t,t-1}, \\
a_{t-1} &= \alpha_{t-1,1} + \alpha_{t-1,2} + \ldots + \alpha_{t-1,t-2} && + (n - \alpha_{t,t-1}), \\
&\cdot \quad \cdot \quad \cdot \\
a_1 &= (n - \alpha_{21}) + (n - \alpha_{31}) + \ldots && + (n - \alpha_{t1}),
\end{aligned} \tag{2.5.2}$$

it follows that the joint distribution of any $s \leqslant t$ scores is given by summing (2.5.1) subject to the restrictions on the s scores imposed by (2.5.2); in particular, if $s = t$ the probability function of the vector of scores, **a**, may be written as

(2.5.3)
$$f(\mathbf{a};\, C(\pi_{ij})) = \sum_{P_n} \prod_{p>r}^{t} \binom{n}{\alpha_{pr}} \pi_{pr}^{\alpha_{pr}} \pi_{rp}^{n-\alpha_{pr}},$$

where $C(\pi_{ij})$ stands for the configuration of preference probabilities π_{ij}, and P_n denotes the restrictions (2.5.2). Huber (1963a) has shown that the a_i are negatively quadrant dependent, i.e., for any values $x_1,\ldots,x_m$, $m \leqslant t$,

$$\Pr\{a_1 < x_1,\ldots,a_m < x_m\} \leqslant \Pr\{a_1 < x_1\}\ldots\Pr\{a_m < x_m\}.$$

If all objects are equivalent, (2.5.3) becomes

$$f(\mathbf{a};\, C(\tfrac{1}{2})) = 2^{-\frac{1}{2}nt(t-1)} \sum_{P_n} \prod_{p>r}^{t} \binom{n}{\alpha_{pr}}.$$

Since $f(\mathbf{a};\, C(\frac{1}{2}))$ must be a symmetric function of the scores, we have established incidentally that

(2.5.4)
$$g(\mathbf{a};\, n) = \sum_{P_n} \prod_{p>r}^{t} \binom{n}{\alpha_{pr}}$$

is symmetric in the a_i. The function $g(\mathbf{a};\, n)$ gives the number of ways the outcome $\mathbf{a}$ can be realized and is closely related to the partition frequencies of §2.3. The latter give the number of permissible partitions of $\frac{1}{2}nt(t-1)$ into t scores a_1, $a_2,\ldots,a_t$, *irrespective* of order, and may be denoted by $G(\mathbf{a};\, n)$. In view of the symmetry of g we have

(2.5.5)
$$G(\mathbf{a};\, n) = \frac{t!}{\prod_k m_k!}\, g(\mathbf{a};\, n),$$

where m_k is the number of scores all of magnitude a_k.

Means, variances, and covariances of scores

If $a_{i\gamma}$ is the score made by A_i in the γth replication we have at once

$$\mathscr{E}(a_{i\gamma}) = \sum_j^{t}{}' \pi_{ij}, \quad \operatorname{var}(a_{i\gamma}) = \sum_j^{t}{}' \pi_{ij}\pi_{ji}.$$

To find the covariance of $a_{i\gamma}$ and $a_{j\gamma}$ $(i \neq j)$, consider the variance of $a_{i\gamma} - a_{j\gamma}$, taking $i=1$ and $j=t$ for convenience. Then

$$\begin{aligned}
\operatorname{var}(a_{1\gamma} - a_{t\gamma}) &= \operatorname{var}\left(\sum_{k=2}^{t} x_{1k\gamma} - \sum_{k=1}^{t-1} x_{tk\gamma}\right) \\
&= \operatorname{var}\left[\sum_{k=2}^{t-1} (x_{1k\gamma} - x_{tk\gamma}) + 2x_{1t\gamma} - 1\right] \\
&= \sum_{k=2}^{t-1} (\pi_{1k}\pi_{k1} + \pi_{tk}\pi_{kt}) + 4\pi_{1t}\pi_{t1} \\
&= \operatorname{var}(a_{1\gamma}) + \operatorname{var}(a_{2\gamma}) + 2\pi_{1t}\pi_{t1}.
\end{aligned}$$

Thus $\quad \operatorname{cov}(a_{1\gamma}, a_{t\gamma}) = -\pi_{1t}\pi_{t1}.$

From the independence of replications it follows that

$$\operatorname{var}(a_1) = n \sum_{k=2}^{t} \pi_{1k}\pi_{k1}, \quad \operatorname{cov}(a_1, a_t) = -n\pi_{1t}\pi_{t1}. \tag{2.5.6}$$

Differences of scores

Of special interest for later work is the distribution of the vector of differences $\mathbf{b}' = (b_1, b_2, \ldots, b_{t-1})$, where

$$b_i = a_i - a_t \quad (i = 1, 2, \ldots, t-1).$$

The variate b_1 has mean

$$\beta_1 = n\left(\sum_{k=2}^{t} \pi_{1k} - \sum_{k=1}^{t-1} \pi_{tk}\right) \tag{2.5.7}$$

and variance

$$\sigma_{b_1 b_1} = n\left[\sum_{k=2}^{t-1} (\pi_{1k}\pi_{k1} + \pi_{tk}\pi_{kt}) + 4\pi_{1t}\pi_{t1}\right]. \tag{2.5.8}$$

Also the covariance of b_1 and b_2 is

$$\sigma_{b_1 b_2} = n\left[\sum_{k=1}^{t-1} \pi_{tk}\pi_{kt} + (\pi_{1t}\pi_{t1} + \pi_{2t}\pi_{t2} - \pi_{12}\pi_{21})\right].$$

Now $b_i = \sum_{\gamma=1}^{n} (a_{i\gamma} - a_{t\gamma}) = \sum_{\gamma} b_{i\gamma}$,

and the $b_{i\gamma}$ have, for any given γ, means, variances and co-variances β_1/n, $\sigma_{b_1b_1}/n$ and $\sigma_{b_1b_2}/n$, respectively. It follows from the independence of replications and the multivariate central limit theorem (e.g., Anderson, 1984) that the limiting distribution as

$n \to \infty$ of $\frac{1}{\sqrt{n}}(\mathbf{b} - \boldsymbol{\beta})$ is multivariate normal $\mathrm{N}(\mathbf{0}, \boldsymbol{\Sigma})$, where $\boldsymbol{\Sigma}$ is the

matrix $\frac{1}{n}(\sigma_{b_ib_j})$.

2.6 A multivariate generalization

Suppose now that the objects $A_1, A_2, \ldots, A_t$ are compared on the basis of two characteristics, say (α, β) (Sen and David, 1968). Then we can define preference frequencies $\alpha_{ij.k}$ $(k = 1, 2, 3, 4)$ and probabilities $\pi_{ij.k}$ corresponding to the four possible outcomes in the n_{ij} comparisons of A_i and A_j $(i \neq j)$:

$$E_{ij}^{(1)}: \alpha_i \to \alpha_j, \beta_i \to \beta_j, \qquad E_{ij}^{(2)}: \alpha_i \to \alpha_j, \beta_i \leftarrow \beta_j,$$
$$E_{ij}^{(3)}: \alpha_i \leftarrow \alpha_j, \beta_i \to \beta_j, \qquad E_{ij}^{(4)}: \alpha_i \leftarrow \alpha_j, \beta_i \leftarrow \beta_j.$$

Note that for all i, j, and k $(i \neq j)$:

$$\pi_{ij.k} = \pi_{ji.5-k} \quad \text{and} \quad \alpha_{ij.k} = \alpha_{ji.5-k}. \tag{2.6.1}$$

The n_{ij} comparisons, assumed independent, result in the multinomial distribution

$$\frac{n_{ij}!}{\prod_{k=1}^{4} \alpha_{ij.k}!} \cdot \prod_{k=1}^{4} \pi_{ij.k}^{\alpha_{ij.k}} \tag{2.6.2}$$

The null hypothesis of equality of the objects with respect to both (α, β) may be expressed as

$$\pi_{ij \cdot k} = \pi_k \quad k = 1, 2, 3, 4, \quad i \neq j = 1, 2, \ldots, t,$$

where by virtue of (2.6.1) we have

$$\pi_1 + \pi_2 = \pi_1 + \pi_3 = \pi_2 + \pi_4 = \pi_3 + \pi_4 = \tfrac{1}{2}. \tag{2.6.3}$$

Equivalently, (2.6.3) may be written as

$$\pi_1 = \pi_4 = \tfrac{1}{4}(1+\theta), \quad \pi_2 = \pi_3 = \tfrac{1}{4}(1-\theta),$$

where θ is an association parameter $(-1 \leqslant \theta \leqslant 1)$. Correspondingly, the null hypothesis may be put in the form

$$\text{(2.6.4)} \quad \begin{aligned} H_0\colon \pi_{ij.k} &= \tfrac{1}{4}(1+\theta) \qquad k=1,4, \\ &= \tfrac{1}{4}(1-\theta) \qquad k=2,3, \text{ for } i<j=1,2,\ldots,t. \end{aligned}$$

Under H_0 the likelihood function of the entire sample is from (2.6.2)

$$\left(\prod_{i<j}^{t} \frac{n_{ij}!}{\prod_{k=1}^{4} \alpha_{ij\cdot k}!}\right) \frac{(1+\theta)^{\alpha_1}(1-\theta)^{N-\alpha_1}}{4^N},$$

where

$$\alpha_1 = \sum_{i<j}^{t} (\alpha_{ij\cdot 1} + \alpha_{ij\cdot 4}) \quad \text{and} \quad N = \sum_{i<j}^{t} n_{ij}.$$

Thus the maximum-likelihood estimator (MLE) of θ is simply

$$\text{(2.6.5)} \qquad \hat{\theta}_N = \frac{2\alpha_1 - N}{N} = \frac{\alpha_1 - \alpha_2}{N}, \quad (\alpha_2 = N - \alpha_1),$$

i.e., $\hat{\theta}_N$ is the difference in the proportions of like and unlike preferences for (α, β).

In order to test H_0 define

$$\begin{aligned} Z_{N,i}^{(1)} &= \sum_j{}' n_{ij}^{-\frac{1}{2}} (\alpha_{ij.1} - \alpha_{ij.4}) \\ & \qquad\qquad\qquad\qquad i \neq j = 1, 2, \ldots, t \\ Z_{N,i}^{(2)} &= \sum_j{}' n_{ij}^{-\frac{1}{2}} (\alpha_{ij.2} - \alpha_{ij.3}). \end{aligned}$$

(If any $n_{ij} = 0$ omit the corresponding term.) We confine ourselves to a statement of the main result. Let

$$\text{(2.6.6)} \qquad \lim_{N\to\infty} \frac{n_{ij}}{N} = \rho_{ij}, \quad \text{where } 0 < \rho_{ij} < 1 \text{ for all } i<j=1,2,\ldots,t.$$

Then for $|\theta| < 1$, under H_0 in (2.6.4), the test-statistic

$$D_N = \frac{N}{t} \sum_{m=1}^{2} \frac{1}{\alpha_m} \sum_{i=1}^{t} (Z_{n,i}^{(m)})^2 \tag{2.6.7}$$

tends to a $\chi^2_{2(t-1)}$ distribution as $N \rightarrow \infty$.

To facilitate extension from 2 to p characteristics, define

$$T_{N,i}^{(1)} = Z_{N,i}^{(1)} + Z_{N,i}^{(2)}, \quad T_{N,i}^{(2)} = Z_{N,i}^{(1)} - Z_{N,i}^{(2)}.$$

Then D_N may be written as

$$D_N = \frac{1}{t(1-\hat{\theta}_N)^2} \sum_{i=1}^{t} [(T_{N,i}^{(1)})^2 - 2\hat{\theta}_N T_{N,i}^{(1)} T_{N,i}^{(2)} + (T_{N,i}^{(2)})^2]. \tag{2.6.8}$$

In the p-dimensional case we need $\frac{1}{2}p(p-1)$ association parameters corresponding to θ. Define the $p \times p$ matrix $\mathbf{\Theta} = (\hat{\theta}_{N,gh})$, $g, h = 1, 2, \ldots, p$, where $\hat{\theta}_{N,gh}$ is the estimate for characteristics g and h, and $\hat{\theta}_{N,gg} = 1$, $g = 1, \ldots, p$. If $\hat{\theta}_N^{gh}$ is the (g, h) element in $\mathbf{\Theta}^{-1}$, then the test statistic is

$$D_{N(p)} = t^{-1} \sum_{g=1}^{p} \sum_{h=1}^{p} \hat{\theta}_N^{gh} \sum_{i=1}^{t} T_{N,i}^{(g)} T_{N,i}^{(h)}.$$

Under the null hypothesis of homogeneity of the t objects, $D_{N(p)}$ tends to a $\chi^2_{p(t-1)}$ distribution as $N \rightarrow \infty$.

Notes

(1) From Table 2.1 we see that the number of $\bar{2}$-partitions of 5 is 4. For t objects the number is the coefficient of x^t in the expansion of

$$1/(1-x)(1-x^3)(1-x^4)(1-x^5)\ldots$$

(Bezembinder, 1981)

(2) The number of distinct score vectors, $a(t)$, has been determined by Narayana and Bent (1964) for $t \leqslant 36$ and the number of nonisomorphic simple Round Robin tournaments, $T(t)$, by Davis (1954) for $t \leqslant 8$. See Moon (1968, p. 87) who lists $T(t)$ for $t \leqslant 12$ and also displays all nonisomorphic tournaments for $t \leqslant 6$, together with their score vectors. Examples: $a(5) = 9$, $T(5) = 12$,

$a(10) = 1486$, $T(10) = 9\,733\,056$. For $t = 5$ the following two non-isomorphic tournaments have the same score vector $[3^2 21^2]$: If an arc joining two nodes has not been drawn, then it is directed from the higher to the lower node (Fig. 2.2).

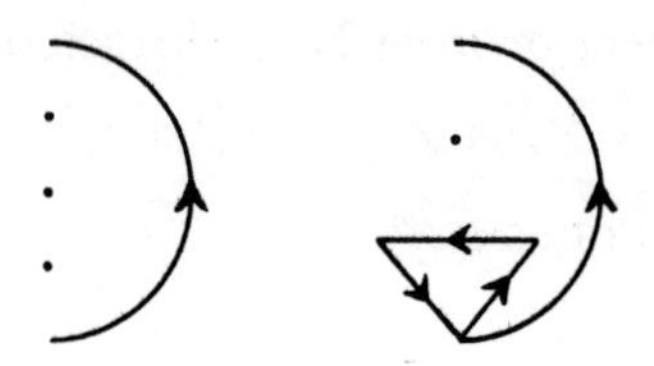

Fig. 2.2

(3) Other mathematical aspects of simple Round Robin tournaments are reviewed in Harary, Norman, and Cartwright (1965, chapter 11) and more fully in Moon (1968).

EXERCISES

2.1 The coefficient of consistence ζ is zero when the number of circular triads c is a maximum.

(Kendall and Babington Smith, 1940)

2.2 If the objects $A_1, A_2, \ldots, A_t$ are equivalent, the mean and variance of c are given by

$$\mu = \frac{1}{4}\binom{t}{3}, \quad \sigma^2 = \frac{3}{16}\binom{t}{3}$$

(Moran, 1947)

2.3 Prove (2.4.4).

2.4 Let d be the difference in the scores of A_1 and A_2 when compared once with each other and $A_3, A_4, \ldots, A_t$. Show that under the assumption H_0 of randomness the probability distribution and characteristic function of d are respectively

$$p(d) = \frac{1}{2^{2t-3}}\left[\binom{2t-4}{t-d-1} + \binom{2t-4}{t-d-3}\right],$$

$$d = -(t-1),\ -(t-2),\ldots,(t-1),$$

and $\phi(u) = (\cos \tfrac{1}{2}u)^{2t-4} \cos u$.

Hence prove (2.4.5).

2.5 For $n = 1$ prove that on H_0

$$\Pr(x_1 \geqslant t-2) = \frac{t^2}{2^{t-1}} - \frac{\binom{t}{2}(t-1)}{2^{2t-4}} + \frac{\binom{t}{3}}{2^{3t-7}}.$$

2.6 Show that on H_0, x_1 is distributed asymptotically ($n \to \infty$) as

$$\tfrac{1}{2}n(t-1) + \tfrac{1}{2}(u_{\max} - \bar{u})\sqrt{(nt)},$$

where $u_{\max}$, $\bar{u}$ are respectively the largest and the mean of t unit normal variates.

2.7 For the set of $(t-1)$ mutually orthogonal contrasts

$$Q_k = \sum_{i=1}^{t} L_{i(k)} d_i \quad (k = 1, 2, \ldots, t-1),$$

where $\sum_{i=1}^{t} L_{i(k)} = 0$ and d_i is defined in (2.4.11), show that the variance of Q_k is

$$S_k = \sum_{i=1}^{t} L_{i(k)}^2$$

and that

$$D_n = \sum_{k=1}^{t-1} Q_k^2 / S_k.$$

2.8 Show that in the non-null case the coefficient of agreement u has mean and variance

$$\mathscr{E}(u) = 1 - \frac{8}{t(t-1)} \sum_{i<j} \pi_{ij}\pi_{ji},$$

$$\operatorname{var}(u) = \frac{64}{n(n-1)t^2(t-1)^2} \sum_{i<j} \{(n-1)[\pi_{ij}\pi_{ji} - 4\pi_{ij}^2\pi_{ji}^2]$$

$$+ 2\pi_{ij}^2\pi_{ji}^2\}.$$

(Ehrenberg, 1952)

2.9 If the integers $a_1, a_2, \ldots, a_t$ sum to $\frac{1}{2}t(t-1)$, they represent a possible outcome of a simple paired-comparison experiment if, and only if,

$$\sum_r a_i \geqslant \tfrac{1}{2}r(r-1) \quad \text{for} \quad r = 1, 2, 3, \ldots, t-1,$$

where $\sum_r a_i$ is the sum over any r of the a_i.

(Landau, 1953; Alway, 1962a)

2.10 If $a_{(1)} \leqslant a_{(2)} \leqslant \ldots \leqslant a_{(t)}$ are the ordered scores of a simple paired-comparison experiment, then

$$\tfrac{1}{2}(i-1) \leqslant a_{(i)} \leqslant \tfrac{1}{2}(t+i-2) \quad i = 1, 2, \ldots, t.$$

[Hint: Show that $a_{(i)} < \frac{1}{2}(i-1)$ leads to a contradiction.]

(Landau, 1953)

2.11 If $(A_1, A_2, \ldots, A_t)$ is a nearest adjoining order in a simple Round Robin tournament, show that $A_1 \to A_2 \to \ldots \to A_t$.
[*or*: For a complete asymmetric digraph every nearest adjoining order is a Hamiltonian path.]

(Remage and Thompson, 1966)

3 NONPARAMETRIC TESTS OF SIGNIFICANCE

3.1 Introductory

This chapter consists of a variety of significance tests on the scores a_i in a balanced paired-comparison experiment. Many of these tests run parallel to well-known tests for differences of treatment means in an analysis of variance. The relevant distribution theory has mostly been derived in §2.4. Here the use of the tests will be illustrated by means of a worked example. Also the necessary tables, other than standard ones, are given.

The reader may wish to convert the tests of significance presented into confidence statements, as can be done in the usual fashion.

3.2 Test of one particular object

Occasionally a certain object, say A_1, is of special interest prior to the paired-comparison experiment. In such a case the experimenter may wish to test the specific null hypothesis

$$H_0' : \pi_{1.} = \tfrac{1}{2}$$

that A_1 is just average against the alternative

$$H_a : \pi_{1.} > \tfrac{1}{2}$$

that it is better than average. Under the hypothesis H_0 of randomness,

$$H_0 : \pi_{ij} = \tfrac{1}{2} \quad \text{for all } i, j\,(i \neq j),$$

the score a_1 of object A_1 is a binomial $b\left(\tfrac{1}{2}, n(t-1)\right)$ variate, and the usual binomial test can be applied to test H_0 against H_a. To test H_0' itself, note that this hypothesis may hold without all π_{1i}'s being equal. Then a_1 is a generalized binomial variate with the same expected value as under H_0, but with a variance that is less than under H_0 by the amount

$$n \sum_{i=2}^{i} (\pi_{1i} - \tfrac{1}{2})^2$$

(see, e.g., Kendall and Stuart, 1977, §5.10). This reduction of variance suggests, as may be shown rigorously (Hoeffding, 1956), that the binomial test of H_0' versus H_a will, if it is not exact, be a conservative test procedure; that is, if the level of significance of the binomial test is α under H_0 it is at most α under H_0'. In this sense the binomial test, appropriate in the first instance as a test of H_0 against H_a, is a safe test of H_0' against H_a. There are strong indications that similar conservative properties apply to the tests developed in the following sections.

The binomial property can also be used, of course, to test H_0' against the alternatives $\pi_{1.} < \frac{1}{2}$ and $\pi_{1.} \neq \frac{1}{2}$. In fact, the conservative property still holds when the more general null hypothesis

$$\pi_{1.} = \pi_0 \qquad (0 < \pi_0 < 1)$$

is tested by treating a_1 as a $b(\pi_0, n(t-1))$ variate. Often the normal approximation to the binomial will give a test of sufficient accuracy.

3.3 Test of equality of two particular objects

Suppose interest is expressed, prior to the performance of the t-object experiment, concerning the existence of a difference between two objects, say A_1 and A_2. After the experiment, we may test

$$H_0' : \pi_{1.} = \pi_{2.}$$

against

$$H_a : \pi_{1.} \neq \pi_{2.}.$$

The distribution of the test criterion $|a_1 - a_2|$ has been developed in §2.4 under the hypothesis of randomness H_0. Hence to test H_0' against H_a we follow the

Test procedure

(1) Choose the desired significance level α.

(2) Find m_c, the smallest integral value of m for which $P_{tnm} = \Pr(|a_1 - a_2| \geqslant m | H_0)$ does not exceed α.

(3) Accept H_a if $|a_1 - a_2|$ exceeds or equals m_c.

To test H'_0 against the one-sided alternative $\pi_{1.} > \pi_{2.}$ use the above test procedure with $\frac{1}{2}P_{tnm}$ in place of P_{tnm} in Step 2 and remove the absolute value sign in Step 3. The critical values of m for one- and two-sided tests when $\alpha = 0{\cdot}01$ or $0{\cdot}05$ are given in Appendix Table 2.

3.4 Test of the top score

After running a paired-comparison experiment one often wishes to know whether the object with the highest score x_1 is significantly better than average. Call this object A_μ and the corresponding mean preference probability $\pi_{\mu.}$. Then the null hypothesis of interest is

$$H'_0 : \pi_{\mu.} = \tfrac{1}{2},$$

which is to be tested against the alternative

$$H_a : \pi_{\mu.} > \tfrac{1}{2}.$$

The discussion of the distribution of the top score x_1 in §2.4 then leads to the

Test procedure

(1) Choose the desired significance level α.

(2) From a table of the cumulative binomial probability distribution find the integer M_β such that

$$t \Pr(a_1 \geqslant M_\beta | H_0) = \beta \leqslant \alpha < t \Pr(a_1 \geqslant M_\beta - 1 | H_0).$$

(3) If the largest score in the experiment exceeds or equals M_β, accept the hypothesis that the corresponding object is better than average.

With obvious modifications the test procedure could be used to test whether the object with the smallest score is below average.

Clearly the test of this section is different in nature from the preceding tests in that no prior interest in any particular objects is postulated. It should also be noted that the specific null hypothesis

H'_0 is here in fact equivalent to the hypothesis of randomness H_0 if the basic probability model is linear. However, in the absence of such linearity $\pi_{\mu.}$ can be $\frac{1}{2}$ without equality of the $\pi_{\mu i}$.

3.5 Overall test of equality

We come now to a test which is the analogue of the F-test for equality of treatment means in an analysis of variance. Such an overall test of equality of the objects can be based on the standardized sum of squares of the scores D_n of §2.4. The null hypothesis under test is

$$H'_0 : \pi_{i.} = \tfrac{1}{2} \qquad \text{all } i,$$

the alternative being that not all the $\pi_{i.}$ are equal. H'_0 is equivalent to H_0 for an underlying linear model.

Test procedure

(1) Choose the desired significance level α.

(2) (a) For small experiments use Appendix Table 1 to find the critical value of the sum of squares of the scores Σa_i^2.

(b) For larger experiments compare

$$D_n = 4\left[\sum_{i=1}^{t} a_i^2 - \tfrac{1}{4}tn^2(t-1)^2\right] \Big/ nt$$

with the upper α significance point of the χ^2-distribution with $t-1$ degrees of freedom.

(3) Reject H'_0 if the observed value of Σa_i^2 or of D_n exceeds or equals the corresponding critical value.

3.6 Least significant difference method

As the name suggests, this method is analogous to carrying out paired t-tests subsequent to finding F significantly large in an overall test of equality of treatment means. In the context of analysis of variance this is the oldest of many multiple comparison tests(*) whose purpose is to answer the question: which treatments

(*) A general account of such tests is given in Miller (1981).

are different from which? In the paired-comparison situation all that needs to be done is to follow up a significant value of Σa_i^2 or D_n by a series of two-sided tests for two particular treatments as in §3.3.

Test procedure

(1)–(3) as in §3.5.

(4) If no significant difference between objects is found in Step 3, accept H_0' and proceed no further; otherwise find the critical value of m_c for a two-sided test on two particular objects as in §3.3 and declare every pair of scores differing by m_c or more to be significantly different.

3.7 Multiple comparison range test

It is well known (e.g. Duncan, 1955) that following up a significant overall F-test by a series of t-tests may lead to quite a high probability of declaring some treatments different when they are not. On the other hand, the method due to Tukey which we parallel in this section suffers perhaps from the opposite fault of tending to make it difficult for true treatment differences to show themselves. However, this approach has the attractive feature that pairs of treatments may be picked out for comparison after inspection of the experimental results, and yet the probability that there will be *any* incorrect declarations of differences can be controlled at the chosen level of significance α.

For paired-comparison experiments the procedure is based on the range of the scores obtained by the t objects. Applying the theory of §2.4 to Tukey's method we obtain the

Test procedure

(1) Choose the desired significance level α.

(2) Find a positive integer $R_{\beta(\alpha)}$, say, such that

$$\text{(3.7.1)} \qquad \Pr[\text{range } a_i \geqslant R_{\beta(\alpha)}] = \beta \leqslant \alpha < \Pr[\text{range } a_i \geqslant R_{\beta(\alpha)} - 1],$$

where the probabilities are calculated under the null hypothesis H_0 of randomness.

(a) If the experiment is small, use Appendix Table 3 to obtain $R_{\beta(\alpha)}$(*)

(b) If the experiment is too large for method (*a*), find the upper α significance point, $W_{t,\alpha}$ say, of the W_t distribution (Pearson and Hartley, 1970, Table 22) and solve

$$W_{t,\alpha} = (2R^* - \tfrac{1}{2})(nt)^{-\frac{1}{2}}$$

for R^*. If R^+, the smallest integer equal to or greater than R^*, is larger than $n(t-1) - \tfrac{1}{2}n$, use equation (3.7.1) to obtain β and $R_{\beta(\alpha)}$; otherwise, set $R_{\beta(\alpha)} = R^+$.

(3) Declare any pairwise difference in scores equal to or greater than $R_{\beta(\alpha)}$ to be significant.

3.8 A method for judging contrasts of scores

Contrasts of scores have been defined in Exercise 2.7. In the analysis of variance the *F*-test has been shown by Scheffé (1953) to provide the basis of a multiple comparison test for contrasts of treatment means. Analogously, one can use the *D*-statistic to judge any and in fact any number of contrasts of scores while controlling the probability of erroneous declarations of significance at a chosen value α. The logic behind this procedure is the same as for the multiple comparison range test. In fact, the test procedure given below can be applied directly to differences of scores, which are simple contrasts, but if interest is confined to differences the earlier method is preferable, being less conservative.

Test procedure

(1)–(3) as in §3.5.

(4) If no significant difference between objects is found in Step 3, accept H_0' and proceed no further; otherwise for any selected contrast

$$\sum_{i=1}^{t} L_i d_i \; (\Sigma L_i = 0) \text{ calculate}$$

(a) $Q^2 = (\Sigma L_i d_i)^2 = 4(\Sigma L_i a_i)^2/(nt)$.

(b) $S = \Sigma L_i^2$.

(c) $SD_{n,c}$.

(*)This table is further discussed by Starks and David (1961).

(5) If $Q^2 \geqslant SD_{n,c}$, declare the contrast significantly different from zero.

3.9 Example

To illustrate the test procedures presented in this chapter, they are applied in turn to data from a paired-comparison experiment run on five brands of carbon paper and described by Fleckenstein, Freund and Jackson (1958). The experiment and analysis were originally carried out according to the method described by Scheffé (1952). For the purpose of the present analyses the 7-point scale used has been contracted by ignoring the degree of preference and assigning ties randomly to one of the two members of each tied pair. The number of times each paper was preferred in the 30 repetitions of the experiment was as follows:

$$a_1 = 66; \quad a_2 = 51; \quad a_3 = 89; \quad a_4 = 24; \quad a_5 = 70.$$

We have $t = 5$, $n = 30$ and arbitrarily take a 5 per cent level of significance throughout. Of course, one would not in actual practice carry out all of the tests on the same body of data.

Test of one particular brand

Suppose that before the carbon-paper experiment, brand 5 had received considerable recommendation because of its low cost and reputed quality. In such a case, the experimenter may be particularly interested in testing whether brand 5 is better than the average of the five brands. The test of $H_0' : \pi_{5.} = \frac{1}{2}$ against $H_a : \pi_{5.} > \frac{1}{2}$ proceeds as follows:

(1) Significance level: 5 per cent.

(2) Since $n(t-1) = 30(4) = 120$, we may use the normal approximation to find the critical value, a_c, for the score of the pre-assigned treatment

$$a_c = 1{\cdot}64[n(t-1)/4]^{\frac{1}{2}} + \tfrac{1}{2}n(t-1) + \tfrac{1}{2} = 69{\cdot}48.$$

(3) $a_5 = 70$.

(4) $a_5 > a_c$. (The actual significance level of a_5 is 0·0412 under H_0'.)

Our conclusion is that brand 5 is better than the average of the five brands.

Test of equality of two particular brands

If brand 4 were less expensive than brand 2, we might suppose that there would be interest expressed prior to the experiment on whether or not brand 2 is actually superior to brand 4. The one-sided test of $H_0 : \pi_{4.} = \pi_{2.}$ against $H_a : \pi_{4.}$ would proceed as follows:

(1) Significance level: 5 per cent.
(2) $1{\cdot}64(nt/2)^{\frac{1}{2}} + 0{\cdot}5 = 1{\cdot}64\sqrt{75} + 0{\cdot}5 = 14{\cdot}7, \quad m_c = 15.$
(3) $a_2 - a_4 = 51 - 24 = 27 > m_c$. We accept the alternative hypothesis H_a that brand 2 is superior to brand 4.

Test of the top score

It is natural for the experimenter to wonder whether the highest score x_1, which in this example is the score of brand 3, is actually significantly larger than average; that is, if $\pi_{\mu.} > \frac{1}{2}$. The test of the null hypothesis against this alternative is as follows:

(1) Significance level: 5 per cent.
(2) From binomial tables (e.g., Harvard University, 1955) we find $n = 120$ and $p = \frac{1}{2}$, that

$$\Pr(a_i \geqslant 74 \mid H_0') = 0{\cdot}033/5,$$

and

$$\Pr(a_i \geqslant 73 \mid H_0') > 0{\cdot}05/5.$$

Therefore, $M_\beta = 74$ and $\beta = 0{\cdot}033$.
(3) $x_1 = a_3 = 89 > M_\beta$. Our conclusion is that brand 3 is significantly better than average.

Overall test of equality

(1) Significance level: 5 per cent.
(2) $D_{n,c} \doteqdot \chi^2{}_{4,0{\cdot}05} = 9{\cdot}488.$
(3) $D_n = 4[\Sigma a_i^2 - \frac{1}{4}tn^2(t-1)^2]/(nt)$
$= 4[20\,354 - 18\,000]/150 = 62.77.$

Since $D_n > D_{n,c}$, a significant difference exists between the scores.

Least significant difference method

(1)–(3) as above.
(4) $1{\cdot}96(\tfrac{1}{2}nt)^{\frac{1}{2}}+\tfrac{1}{2}=1{\cdot}96\sqrt{75}+\tfrac{1}{2}=17{\cdot}47.$

The critical difference is $m_c = 18$, giving the ordering

a_4	a_2	a_1	a_5	a_3
24	51	66	70	89 .

Any two brands whose scores are not underlined by the same line may be considered distinguishably different.

Multiple comparison range test

(1) Significance level: 5 per cent.
(2) $W_{5,0{\cdot}05}=3{\cdot}86$ (obtained from Table 22, Pearson and Hartley, 1970).

$$R^* = \tfrac{1}{2}W_{5,0{\cdot}05}\sqrt{(nt)}+\tfrac{1}{4}=3{\cdot}86\sqrt{(150/4)}+\tfrac{1}{4}=23{\cdot}887,$$

$$R^+ = 24 < n(t-1)-\tfrac{1}{2}n = 105,$$

$$\beta \doteqdot \Pr[W_t \geqslant (2R-\tfrac{1}{2})(nt)^{-\frac{1}{2}}] = \Pr[W_5 \geqslant 3{\cdot}878] = 0{\cdot}048.$$

(From Table 23, *loc. cit.*)

(3)

a_4	a_2	a_1	a_5	a_3
24	51	66	70	89

Any two brands whose scores are not underlined by the same line in the above step may be considered distinguishably different.

Method of contrasts

(1)–(3) as for overall test of equality.
(4a) The separation into groups can be accomplished in this case by computing the squares of four contrasts

$$Q_1^2 = 4(a_3-a_2)^2/nt = 4(89-51)^2/150 = 38{\cdot}5,$$

$$Q_2^2 = 4(a_3-a_1)^2/nt = 4(89-66)^2/150 = 14{\cdot}1,$$

$$Q_3^2 = 4(a_5-a_2)^2/nt = 4(70-51)^2/150 = 9{\cdot}6,$$

$$Q_4^2 = 4(a_2-a_4)^2/nt = 4(51-24)^2/150 = 19{\cdot}4.$$

(4b) For each contrast

$$S=\sum_{i=1}^{2} L2_i = 1^2+1^2=2.$$

(4c) $SD_{n,c}=2(9{\cdot}488)=18{\cdot}976.$

(5) Since

$$Q_1^2 > Q_4^2 > SD_{n,c} > Q_2^2 > Q_3^2$$

the resulting ordering is, with the usual interpretation,

a_4	a_2	a_1	a_5	a_3
24	51	66	70	89

More general contrasts can be readily examined as explained in §3.8.

EXERCISES

3.1 In the example of §3.9 determine by the method of contrasts whether a_5 differs significantly, at the 5 per cent level, from the *mean* of a_1 and a_2.

3.2 Show that the multiple comparison range test may be used to declare significant at level α any contrast $\sum_{i=1}^{t} c_i a_i$ $(\Sigma c_i = 0)$ which exceeds or equals

$$\tfrac{1}{2}R_{\beta(\alpha)}\Sigma|c_i|,$$

$R_{\beta(\alpha)}$ being defined in (3.7.1).

[This is the analogue of Tukey's procedure for contrasts. See Scheffé (1953).]

4 THE LINEAR MODEL

4.1 A general approach

The linear model was defined in §1.3. Before treating in detail some of its important special cases we consider a general approach used by Noether (1960). The first problem investigated by him is the estimation of the merits V_i $(i = 1, 2, \ldots, t)$. Since the origin of the linear scale is arbitrary it is convenient to introduce the restriction

$$\sum_{i=1}^{t} V_i = 0, \tag{4.1.1}$$

which permits the unique estimation of the V_i. (Another possibility is to set $V_{(1)}$, the smallest merit, equal to zero, but (4.1.1) has some advantages due to its symmetry.) Let

$$\delta_{ij} = V_i - V_j, \qquad (j = 1, 2, \ldots, t;\ j \neq i) \tag{4.1.2}$$

so that the preference probability π_{ij} is given by

$$\pi_{ij} = H(\delta_{ij}), \tag{4.1.3}$$

H being the cdf pertaining to the linear model chosen. Summing over all j other than $j = i$, we have

$$\Sigma' \delta_{ij} = (t-1) V_i - \Sigma' V_j,$$

or

$$V_i = \frac{1}{t} \sum_j{}' \delta_{ij}. \tag{4.1.4}$$

This suggests the following method of finding estimates v_i of the V_i for a balanced paired-comparison experiment with n replications. Let $p_{ij} = \alpha_{ij}/n$ be the proportion of preferences for object A_i over A_j. As in (4.1.3) define d_{ij} by

(4.1.5) $$H(d_{ij}) = p_{ij}.$$

(4.1.6) Then $$d_{ji} = -d_{ij}.$$

A difficulty arises at this point if $p_{ij} = 0$ or 1 *and* if H is the cdf of a variate with infinite range; to obtain a finite value for d_{ij} one possibility is to replace the observed value of p_{ij} by $1/(2n)$ or $1 - 1/(2n)$, respectively.

In general, it is impossible to satisfy the relations $v_i - v_j = d_{ij}$, corresponding to (4.1.2), since there are more equations than unknowns. However, these relations can be satisfied "on the average" in the sense of (4.1.4) by setting

(4.1.7.) $$v_i = \frac{1}{t} \sum_j{}' d_{ij}.$$

The v_i obtained in this way are (unweighted) least square estimates of the V_i, i.e. they minimize

(4.1.8.) $$S = \sum_{\substack{i,j \\ i \neq j}} (d_{ij} - V_i + V_j)^2;$$

for $$\frac{\partial S}{\partial V_i} = -2\sum_j{}' (d_{ij} - V_i + V_j) + 2\sum_j{}' (d_{ji} - V_j + V_i)$$

$$= -4\sum_j{}' (d_{ij} - \delta_{ij}) \quad \text{by (4.1.2) and (4.1.6),}$$

$$= -4\left(\sum_j{}' d_{ij} - tV_i\right) \quad \text{by (4.1.4),}$$

which gives (4.1.7) as the solution of $\partial S / \partial V_i = 0$.

Clearly, other estimates of the V_i are possible, but the simple estimates (4.1.7) have the advantage that the method of computation remains the same whatever the assumed form of H may be. This would not in general be the case for estimates based on the methods of maximum likelihood, minimum chi square, etc. All we need is a table or algorithm for the function H, or better still for H^{-1}, so that given p_{ij} we obtain d_{ij} by (4.1.5) as

$$d_{ij} = H^{-1}(p_{ij}).$$

It may be noted here that the minimization of S in (4.1.8) can be extended to unbalanced experiments (Gulliksen, 1956; Kaiser and Serlin, 1978). This may be viewed as a special case of the method of (unweighted) least squares.

Thurstone–Mosteller model

Here the d_{ij}, and hence the v_i, are immediately obtainable from tables or computer programs giving the unit normal deviate exceeded with probability $p_{ji} = 1 - p_{ij}$.

Even simpler is the use of "rankits" as suggested by Bliss, Greenwood, and White (1956), who convert p_{ij} into the expected value of the $(n - \alpha_{ij} + 1)$th largest member of a random sample of $n + 1$ from a unit normal population. The rankits (or normal scores) are approximately equal to $H^{-1}(p_{ij})$ and can be read off at once for $n \leqslant 100$ from, e.g., Table 9, Pearson and Hartley (1972). There are no complications for $p_{ij} = 0$ or 1.

Bradley–Terry model

We find from (1.3.10) that

$$p_{ij} \equiv H(d_{ij}) = \tfrac{1}{2}[1 + \tanh(\tfrac{1}{2}d_{ij})],$$

and hence that

$$d_{ij} = \log(p_{ij}/p_{ji}).$$

With obvious notation were therefore have

$$v_i = \log \prod_j{}' \, (\alpha_{ij}/\alpha_{ji}) \qquad (4.1.9.)$$

However, it is to be noted that the v_i estimate parameters V'_i (say) for which

$$\Sigma V'_i = \Sigma \log \pi'_i = 0,$$

i.e. the parameters π'_i do not satisfy $\Sigma \pi'_i = 1$, the restriction used by Bradley and Terry, but

$$\pi'_1 \pi'_2 \ldots \pi'_t = 1.$$

The π'_i need only be scaled down to give the π_i

$$\pi_i = \pi'_i / \Sigma \pi'_i,$$

which shows that

$$\pi_{ij} \equiv \frac{\pi_i}{\pi_i + \pi_j} = \frac{\pi'_i}{\pi'_i + \pi'_j}.$$

The uniform distribution

This important case arises if we take $H(x) = x + \frac{1}{2}$ $(-\frac{1}{2} \leqslant x \leqslant \frac{1}{2})$. Then $d_{ij} = p_{ij} - \frac{1}{2}$, and

$$V_i = \frac{1}{t} \sum_j{}' (p_{ij} - \tfrac{1}{2}) \tag{4.1.10}$$

$$= \frac{1}{tn} [a_i - \tfrac{1}{2} n(t-1)],$$

i.e. the v_i are linear functions of the scores a_i, and a ranking of the objects made according to the a_i is the same as one according to the v_i. Taking expectations we have

$$V_i = \frac{t-1}{t} (\pi_{i.} - \tfrac{1}{2}).\text{(*)}$$

Thus, from the point of view of estimating the merits, assuming a uniform distribution is equivalent to working with the scores as in the preceding two chapters. Of course, the present assumptions are stronger and determine the π_{ij} as functions of the $\pi_{i.}$ since

$$\pi_{ij} \equiv H(V_i - V_j) = V_i - V_j + \tfrac{1}{2}$$

$$= \frac{t-1}{t} (\pi_{i.} - \pi_{j.}) + \tfrac{1}{2}. \tag{4.1.11}$$

Noether goes on to point out that to the extent that all functions $H(x)$ are approximately linear in the neighbourhood of $x = 0$, it is to be expected that for sets of V_i which do not differ much among themselves estimates will be approximately equivalent to those for the uniform distribution whatever function $H(x)$ is actually used. In fact, if n is not too small, we have approximately

(*) Here $\pi_{i.} = \frac{1}{t-1} \sum_j{}' \pi_{ij}$, not $\frac{1}{t} \sum_{j=1}^{t} \pi_{ij}$ $(\pi_{ii} = \frac{1}{2})$ as used by Noether.

$$p_{ij} - \pi_{ij} = H(d_{ij}) - H(\delta_{ij}) \doteqdot (d_{ij} - \delta_{ij})\, h(\delta_{ij}),$$

where $h(x) = H'(x)$. It follows that if the V_i do not differ much, a good approximation is usually given by

$$d_{ij} \doteqdot (p_{ij} - \tfrac{1}{2})/h(0)$$

or

$$v_i \doteqdot \frac{1}{nt\,h(0)}\,[a_i - \tfrac{1}{2}n(t-1)], \tag{4.1.12}$$

which is simply a multiple of (4.1.10).

Test of the hypothesis $V_1 = V_2 = \ldots = V_t$

Before using a given set of estimates v_i it is desirable to establish that the corresponding V_i are unequal. This can be done by testing (and rejecting) the hypothesis $V_1 = V_2 = \ldots = V_t$ by means of the statistic

$$B_h = 4nt\,h^2(0) \sum_{i=1}^{t} v_i^2, \tag{4.1.13}$$

where the subscript h indicates that the v_i have been computed by (4.1.7) using the function $h(x)$. Since (4.1.12) holds asymptotically if the null hypothesis is true,

$$B_h \sim \frac{4}{nt} \sum_{i=1}^{t} [a_i - \tfrac{1}{2}n(t-1)]^2,$$

so that B_h has a χ^2 distribution with $t-1$ DF, being asymptotically equivalent to the D_n-statistic of §2.4. This means that if we are interested only in an asymptotic test of the null hypothesis $V_1 = V_2 = \ldots = V_t$, it makes no difference which law $H(x)$ is chosen.

4.2 The angular transformation

An objection which may be raised against the foregoing general approach is that the sum of squares S minimized in (4.1.8) is an unweighted sum. For example, in the case of the Thurstone–

Mosteller model the d_{ij}, although obtainable from normal tables, are not unit normal deviates since they can assume only $n+1$ distinct values corresponding to the $n+1$ values of p_{ij}. Moreover, the variance of each d_{ij} depends on the true value π_{ij}. If n is sufficiently large we have

$$\begin{aligned}\mathrm{var}(d_{ij}) &= \mathrm{var}[H^{-1}(p_{ij})] \\ &\doteqdot \mathrm{var}(p_{ij})\left\{\frac{d}{d\pi_{ij}}[H^{-1}(\pi_{ij})]\right\}^2 \\ &= \frac{\pi_{ij}\pi_{ji}}{n}\cdot\frac{1}{[h(\pi_{ij})]^2} \\ &= \frac{1}{W_{ij}} \quad \text{(say).}\end{aligned}$$

Thus we should minimize

$$\sum_{i \neq j} W_{ji}(d_{ji} - V_i + V_j)^2,$$

but the weights W_{ij} are unknown, so that a laborious iterative solution would be needed (see Bock and Jones, 1968, p. 124). The purpose of the angular transformation is to avoid these difficulties. Setting

$$d_{ij} = 2\sin^{-1}\sqrt{p_{ij}} - \tfrac{1}{2}\pi = \sin^{-1}(2p_{ij} - 1), \tag{4.2.1}$$

we have approximately, for large n,

$$\mathrm{var}(d_{ij}) = 1/n.$$

If need be, the variance-stabilizing properties of the transformation can be further improved for small n by replacing p_{ij} by

$$(\alpha_{ij} + 3/8)/(n + 3/4).$$

Note that (4.2.1) corresponds to taking

$$p_{ij} = H(d_{ij}) = \tfrac{1}{2}(1 + \sin d_{ij}). \tag{4.2.2}$$

We might therefore have treated the angular transformation in §4.1. The only new feature is that the unweighted sum of squares S is here approximately the same as the weighted sum of squares.

Moreover, in all but the lower and upper 5 per cent tail areas the angular transformation closely parallels the normal except for a scale factor.

4.3 The Bradley–Terry model—estimation[*]

Zermelo (1929) and Bradley and Terry (1952) use the method of maximum likelihood to estimate the t "ratings" $\pi_i (\pi_i \geqslant 0, \Sigma \pi_i = 1)$ occurring in their model

$$\pi_{ij} = \pi_i/(\pi_i + \pi_j) \qquad (i, j = 1, 2, \ldots, t; i \neq j).$$

In view of the independence of all comparisons the probability of observing α_{ij} preferences for A_i in its n comparisons with A_j is

$$\binom{n}{\alpha_{ij}} \left(\frac{\pi_i}{\pi_i + \pi_j}\right)^{\alpha_{ij}} \left(\frac{\pi_j}{\pi_i + \pi_j}\right)^{\alpha_{ji}} \qquad (\alpha_{ij} = 0, 1, \ldots, n).$$

The likelihood function L is the product of such probabilities for any $\binom{t}{2}$ independent pairings and may be expressed in terms of the scores a_i as

$$L = C \frac{\prod_{i=1}^{t} \pi_i^{a_i}}{\prod_{i<j} (\pi_i + \pi_j)^n}, \tag{4.3.1}$$

where $C = \prod_{i<j} \binom{n}{\alpha_{ij}}$ is independent of the π_i. It follows that the a_i are sufficient statistics for the π_i. In fact, it can be shown (Bühlmann and Huber, 1963) that for any preference matrix with $\pi_{ij} > 0$, all $i \neq j$, the scores are sufficient statistics only under the Bradley–Terry model. See also Huber (1963b).

Differentiating $\log L$ with respect to π_i we obtain the maximum likelihood (ML) estimates p_i of the π_i from any $t-1$ of the equations

$$\frac{a_i}{p_i} - n \sum_j{}' \frac{1}{p_i + p_j} = 0, \tag{4.3.2}$$

[*]See Note 2 to Chapter 1 (p. 13).

taken together with

$$\Sigma p_i = 1.$$

We see that the p_i are functions of the a_i and do not involve the individual α_{ij}, in contrast to the estimates (4.1.9). To solve for the p_i it is convenient to write (4.3.2) in the form

$$(4.3.3) \qquad p_i = \frac{a_i}{n \sum_j{}' (p_i + p_j)^{-1}}.$$

Then starting with a set of trial solutions $(p_1^{(0)}, p_2^{(0)}, \ldots, p_t^{(0)})$ we can obtain $p_i^{(1)}$ $(i = 1, 2, \ldots, t)$ from

$$(4.3.4) \qquad p_i^{(1)} = \frac{a_i}{n} \left[\frac{1}{p_i^{(0)} + p_1^{(1)}} + \ldots + \frac{1}{p_i^{(0)} + p_{i-1}^{(1)}} + \frac{1}{p_i^{(0)} + p_{i+1}^{(0)}} + \ldots + \frac{1}{p_i^{(0)} + p_t^{(0)}} \right]^{-1}$$

continuing the iterative process until agreement between $p_i^{(r+1)}$ and $p_i^{(r)}$ is sufficiently close. Using this method, Bradley and Terry (1952) and Bradley (1954b) have tabulated the p_i to two decimal places for all possible outcomes of experiments of size (t, n) up to (3, 10), (4, 8), and (5, 5), cf. Table 4.1, p. 68.(*) Dykstra (1956) points out that the convergence of the iterative procedure is slow so that it is important to get starting values before applying (4.3.4) to cases outside the range of the tables. For this purpose he recommends that, to begin with, we take all p_j $(j \neq i)$ to be equal, so that

$$p_j = (1 - p_i)/(t - 1),$$

giving p_i from (4.3.3) as

$$(4.3.5) \qquad p_i = a_i/[n(t-1)^2 - a_i(t-2)].$$

This approximation to the ML estimate of π_i may be further improved by replacing p_i in (4.3.5) by $p_i - k_i$, where k_i is an empirical correction factor proposed by Dykstra, viz.,

$$(4.3.6) \quad k_i = \frac{[(t-1)R - N^2]/t + a_i[n(t-1) - a_i]}{(t-1)R - N^2 + \Sigma a_i[n(t-1) - a_i]} \left[\left(\sum_i p_i - 1 \right) \right],$$

(*) If $A_i \rightarrow A_j$ these authors assign rank 1 to A_i and rank 2 to A_j. The resulting rank totals Σr_i are related to the scores a_i by $\Sigma r_i = 2n(t-1) - a_i$.

where $R = \Sigma a_i^2$ and $N = \frac{1}{2}nt(t-1)$.

Note that the p_i of (4.3.5) do not add up to unity, in general. In fact, the iterates $p_i^{(r)}$ also do not add to one for $r > 0$, but to normalize them at each stage of the iteration is not necessary since we are really concerned with their ratios and not their actual values (cf. Ford, 1957).

Zermelo (1929) and Ford make the interesting comment that the ranking achieved by the ML estimates is the same as that of the scores; for if $a_i > a_j$, then at a solution of (4.3.2)

$$0 < n\left(\sum_k{}' \frac{p_i}{p_i + p_k} - \sum_k{}' \frac{p_j}{p_j + p_k}\right)$$

$$= n(p_i - p_j)\sum_k \frac{p_k}{(p_i + p_k)(p_j + p_k)},$$

so that $p_i > p_j$.

Unequal repetitions

If the number of comparisons of A_i with A_j is not constant but equal to n_{ij} the last result may, of course, break down. However, the ML estimation of the p_i goes through with very little change: we need merely replace (4.3.3) by

$$p_i = \frac{a_i}{\sum_j n_{ij}(p_i + p_j)^{-1}}.$$

The convergence of the iterative process has been established by Zermelo and Ford in this general case, provided the following assumption holds:

"In every possible partition of the objects into two non-empty subsets, some object in the second set has been preferred at least once to some object in the first set."

This assumption is a requirement that one might reasonably expect. For if it is violated then there exist two subsets S_1 and S_2 with either no interset comparisons, or with all interset comparisons favouring S_1. In the former case one can obviously not rank any object in S_1 against any object in S_2. In the latter case, which applies also to the balanced experiment with equal n_{ij}, the maximizing values of the π_i for objects in S_2 must turn out to be zero, since otherwise we could increase the likelihood function

$$L = \frac{C \prod_{i=1}^{t} \pi_i^{a_i}}{\prod_{i<j} (\pi_i + \pi_j)^{n_{ij}}} \tag{4.3.7}$$

by multiplying these π_i by some common factor <1 and the remaining π_i by some common factor >1. These factors may be chosen to preserve a sum of 1. The factors of L involving i, j, both in S_1 or both in S_2, remain unchanged by this operation. However, the factors involving i in S_1 and j in S_2 increase, thereby increasing L. But if the π_i for objects in S_2 must all be zero, we cannot hope to compare individual members of S_2 by this process. Of course, we may treat these objects completely separately from those in S_1.

As Dykstra (1960) points out, results for unequal n_{ij} are useful also in designed experiments where only a proportion of all possible comparisons are made in order to reduce the size of the experiment. In this case $n_{ij} = 1$ or 0 according as A_i is compared with A_j or not. There are other situations where the n_{ij} are unequal by design (see Chapter 5).

Precision of the p_i

Standard large-sample theory may be used to obtain the asymptotic distribution of $p_1, \ldots, p_t$ (Bradley, 1955; Dykstra, 1960; Bradley, 1984). Let

$$N = \sum_{i<j} n_{ij}, \quad m_{ij} = \frac{n_{ij}}{N} \quad i, j = 1, 2, \ldots, t; i \neq j,$$

$$\lambda_{ii} = \frac{1}{\pi_i} \sum_j{}' \frac{m_{ij}\pi_j}{(\pi_i + \pi_j)^2}, \quad \lambda_{ij} = \frac{-m_{ij}}{(\pi_i + \pi_j)^2}. \tag{4.3.8}$$

Then $\sqrt{N}(p_1 - \pi_1), \ldots, \sqrt{N}(p_t - \pi_t)$ have, as $N \to \infty$, the singular multivariate normal distribution of $t-1$ dimensions in t-dimensional space with zero means and dispersion matrix $\mathbf{\Sigma} = (\sigma_{ij})$ defined by

(4.3.9)
$$\sigma_{ij} = \frac{\text{cofactor of } \lambda_{ij} \text{ in } \begin{vmatrix} \mathbf{\Lambda} & \mathbf{1} \\ \mathbf{1}' & 0 \end{vmatrix}}{\begin{vmatrix} \mathbf{\Lambda} & \mathbf{1} \\ \mathbf{1}' & 0 \end{vmatrix}},$$

where $\mathbf{\Lambda} = (\lambda_{ij})$ and $\mathbf{1}$ is the column vector of t 1's. In practice, σ_{ij} must be estimated by replacing π_i by p_i in (4.3.8).

Since $\log p_i$ is a location estimate for the ith object A_i, the distribution of the $\log p_i$ is also of interest. It will be seen that the $\sqrt{N}(\log p_i - \log \pi_i)$, $i = 1, 2, \ldots, t$, have the same singular multivariate normal distribution, except that $\mathbf{\Sigma}$ has to be replaced by $\mathbf{D\Sigma D}$, where $\mathbf{D}$ is a diagonal matrix with ith element $1/\pi_i$.

Confidence intervals for the π_i

Large-sample $(1-\alpha)$-confidence regions for any vector π^* consisting of t^* of the parameters $\pi_1, \ldots, \pi_t$ $(t^* < t)$ are given by the ellipsoidal region of the t^*-dimensional parameter space for which

(4.3.10)
$$N(\boldsymbol{\pi}^* - \boldsymbol{p}^*)' \hat{\mathbf{\Sigma}}^{*-1} (\boldsymbol{\pi}^* - \boldsymbol{p}^*) \leqslant \chi^2_{t^*, \alpha}.$$

Bayesian methods

A Bayesian approach to paired comparisons was introduced by Davidson and Solomon (1973). Corresponding to the likelihood function

$$L(\boldsymbol{A} | \boldsymbol{\pi}) = \prod_{i<j} \binom{n_{ij}}{\alpha_{ij}} \frac{\pi_i^{\alpha_{ij}} \pi_j^{\alpha_{ji}}}{(\pi_i + \pi_j)^{n_{ij}}},$$

where $\boldsymbol{A} = (\alpha_{ij})$, the authors assume the following conjugate prior density for $\boldsymbol{\pi} = (\pi_1, \ldots, \pi_t)'$:

$$f(\boldsymbol{\pi}; \boldsymbol{A}^0, \boldsymbol{N}^0) = K(\boldsymbol{A}^0, \boldsymbol{N}^0) \prod_{i<j} \frac{\pi_i^{\alpha_{ij}^0} \pi_j^{\alpha_{ji}^0}}{(\pi_i + \pi_j)^{n_{ij}^0}}.$$

Here $\boldsymbol{A}^0 = \{\alpha_{ij}^0\}$ and $\boldsymbol{N}_0^0 = \{n_{ij}^0\}$ $(i, j = 1, 2, \ldots, t)$ are matrices of prior parameters with $\alpha_{ii}^0 = n_{ii}^0 = 0$, $n_{ij}^0 = n_{ji}^0$, and $\alpha_{ij}^0 + \alpha_{ji}^0 = n_{ij}^0$. The prior density represents the belief that A_i will win α^0_{ij} of n^0_{ij} conceptual encounters of A_i and A_j.

The posterior density of $\boldsymbol{\pi}$ is now simply $f(\boldsymbol{\pi}; \mathbf{A}^1, \mathbf{N}^1)$, where

$$\mathbf{A}^1 = \mathbf{A}^0 + \mathbf{A} \quad \text{and} \quad \mathbf{N}^1 = \mathbf{N}^0 + \mathbf{N},$$

and may alternatively be expressed as (cf. (4.3.7))

$$f(\boldsymbol{\pi}; \mathbf{A}^1, \mathbf{N}^1) = \frac{C \prod_{i=1}^{t} \pi_i^{a_i^1}}{\prod_{i<j} (\pi_i + \pi_j)^{n_{ij}^1}}, \tag{4.3.11}$$

where C does not depend on $\boldsymbol{\pi}$, and $a_i^1 = \sum_{j=1}^{t} \alpha_{ij}^1$ may be called the posterior score of A_i.

It is now clear that the estimation of $\boldsymbol{\pi}$ by maximizing (4.3.11) requires exactly the same iterative procedure as the estimation of $\boldsymbol{\pi}$ in (4.3.7). In addition to this posterior mode estimator, Davidson and Solomon (1973) examine also the less convenient posterior mean estimator and show that under posterior balance ($n_{ij}^1 = n^1$, all $i \neq j$) both estimators rank the A_i in the same order as do the posterior scores a_i^1.

An alternative Bayesian approach due to Leonard (1977) replaces the conjugate prior by a multivariate normal prior of $\log \pi_1, \ldots, \log \pi_t$. The parameters of this prior can sometimes be estimated from previous data, especially if they can be assumed to have a simple structure. See also Chen and Smith (1984) for non-iterative estimation of $\boldsymbol{\pi}$.

Davidson and Solomon (1973) treat also the product of binomial probabilities models of Chapter 2. The conjugate priors are correspondingly products of beta densities. By imposing restrictions on these densities suggested by the Bradley–Terry model, Lancaster and Quade (1983) obtain a model allowing for correlations between comparisons by the same judge of the same pair of objects.

4.4 The Bradley–Terry model—tests of hypotheses

Under the Bradley–Terry model, the null hypothesis of equality among the t objects A_i becomes

$$H_0 : \pi_i = 1/t \qquad (i = 1, 2, \ldots, t).$$

By (4.3.1) the corresponding likelihood for the case of equal n_{ij} is

$$C . 2^{-\frac{1}{2}nt(t-1)}.$$

A test of H_0 against the general alternative of unequal π_i may be based on the likelihood ratio

$$\lambda = \frac{\prod_{i<j} (p_i + p_j)^n}{2^{\frac{1}{2}nt(t-1)} \prod_{i=1}^{t} p_i^{a_i}} \tag{4.4.1}$$

Bradley and Terry (1952) use instead a monotonic function of λ, viz.

$$B_1 = n \sum_{i<j} \log_{10}(p_i + p_j) - \Sigma a_i \log_{10} p_i. \tag{4.4.2}$$

When the ML estimates p_i have been found, B_1 may be computed. If this is done for all possible outcomes of an experiment, the probability distribution of B_1 can be constructed from a knowledge of the probabilities of the individual outcomes, in the same way as the probability distribution of $T_n = \Sigma(a_i - \bar{a})^2$ was obtained in §2.4. The situation is illustrated in Table 4.1 for the case $t = 4$, $n = 2$. In our previous notation the first entry is the partition [6420]. Dashes indicate that the corresponding p_i are indeterminate for the reasons given near (4.3.7). The P column gives the probability with which B_1 is equalled or exceeded when H_0 is true. The final column, which we have added, gives the values of T_2 (this is also the sum of squares of the Σr_i). While the ordering of the rank sums was made according to increasing B_1 values, it corresponds in the present case exactly to decreasing values of T_2. Thus the two statistics are here completely equivalent *as tests of* H_0. This is not always so, but the discrepancies, if any, are slight. Of course, in the range of the Bradley–Terry tables one need only locate a particular rank ordering (such as 7, 7, 11, 11 in Table 4.1) to see at once what the corresponding P value is (·0290). For values of n outside the range of the tables when $t \leqslant 5$ and for all but the smallest (t, n) combinations for $t > 5$ one may take

$$-2 \log \lambda = nt(t-1) \log 2 - 2B_1 \log 10 \tag{4.4.3}$$

Table 4.1 Possible rank totals in the case $t=4$, $n=2$, and the associated p_i, B_1, and P

From Bradley and Terry (1952) by permission of the authors and the Editor of *Biometrika*

Σr_1	Σr_2	Σr_3	Σr_4	p_1	p_2	p_3	p_4	B_1	P	T_2
6	8	10	12	1	—	—	—	0·000	·0058	20
6	8	11	11	1	—	—	—			
6	9	9	12	1	—	—	—	0·602	·0232	18
7	7	10	12	·50	·50	—	—			
7	7	11	11	·50	·50	—	—	1·204	·0290	16
6	9	10	11	1	—	—	—	1·498	·0994	14
7	8	9	12	·59	·28	·13	—			
6	10	10	10	1	—	—	—	1·806	·1190	12
8	8	8	12	·33	·33	·33	—			
7	8	10	11	·60	·29	·08	·04	2·359	·2245	10
7	9	9	11	·62	·17	·17	·05	2·631	·3065	8
7	9	10	10	·62	·18	·10	·10	2·898	·5643	6
8	8	9	11	·37	·37	·21	·06			
8	8	10	10	·38	·38	·12	·12	3·158	·6639	4
8	9	9	10	·40	·23	·23	·14	3·389	·9627	2
9	9	9	9	·25	·25	·25	·25	3·612	1·0000	0

to be distributed approximately as χ^2 with $t-1$ DF when the null hypothesis is true. Again, the use of the T_n statistic of §2.1 should lead to very similar results when interest centres on a test of H_0. The asympotic equivalence, under the Bradley–Terry model, of tests based on B_1 and T_n has in fact been demonstrated by Bradley (1955) in the non-null as well as the null case. He takes the alternative hypothesis as

$$H_a : \pi_i = \frac{1}{t} + \frac{\delta_{in}}{\sqrt{n}} \qquad (i = 1, 2, \ldots, t; \sum_i \delta_{in} = 0),$$

where δ_{in} represents a sequence of constants converging to δ_i as $n \to \infty$. As is usual in the derivation of limiting power functions, H_a is actually a sequence of alternatives approaching H_0 with increasing n. The limiting distribution as $n \to \infty$ of $-2 \log \lambda$ and of

$$D_n = \frac{4}{nt} \Sigma(a_i - \bar{a})^2$$

is a non-central χ^2 distribution with $t-1$ DF and non-centrality parameter

$$\lambda' = \tfrac{1}{4}t^3\Sigma\,\delta_i^2.$$

Thus for large n an adequate approximation to the power is obtained by asssuming the non-central χ^2 distribution and taking λ' to be

$$\lambda' = \tfrac{1}{4}nt^3\Sigma(\pi_i - 1/t)^2.$$

For given values of π_i it is therefore possible to apply the non-central χ^2 distribution in standard fashion in order to obtain the smallest value of n guaranteeing with a prescribed probability β that H_0 will be rejected by a test at significance level α.

Example

$\alpha = 0{\cdot}05$, $\beta = 0{\cdot}90$, $t = 4$; $\pi_1 = 0{\cdot}4$, $\pi_2 = 0{\cdot}3$, $\pi_3 = 0{\cdot}2$, $\pi_4 = 0{\cdot}1$. We find $\lambda' = 0{\cdot}8n$. Entering the charts of Pearson and Hartley (1951) with $\alpha = 0{\cdot}05$, $\beta = 0{\cdot}90$, $v_1 = t-1 = 3$, $v_2 = \infty$, we see that $\phi \equiv \sqrt{(\lambda'/t)} = 1{\cdot}88$. Hence the smallest satisfactory value of n is 18.

Bradley uses this approach to compare the asymptotic power functions of the present test and of what he terms the multi-binomial test. The latter consists in pooling the $\tfrac{1}{2}t(t-1)$ independent binomial tests which can be made in a preference table by means of the statistic

$$4n\sum_{i<j}(p_{ij} - \tfrac{1}{2})^2$$

which, under H_a, has asymptotically a non-central χ^2- distribution with $\tfrac{1}{2}t(t-1)$ DF and non-centrality parameter λ'. As might be anticipated, the multi-binomial test is distinctly inferior here. The test is, however, of interest in cases where a linear model does not apply.

It should be noted that in view of the essential equivalence of all linear models in the vicinity of the null hypothesis we are once again (cf. §4.1) led to the conclusion that the simple D_n-statistic provides the appropriate large-sample test of H_0.

The asymptotic power function of the D_n-test may also be compared with that of the analysis of variance F-test for a balanced incomplete block design with two *measurements* per block. Under the assumptions of a linear model, van Elteren and Noether (1959) obtain the asymptotic relative efficiency as

$$8\sigma^2[\int h^2(x)\,dx]^2,$$

which is independent of t and reduces to $2/\pi$ when $h(x)$ is normal with variance σ^2.

Combination of experiments

The use of the test (4.4.3) assumes that the same ratings π_i apply throughout the entire experiment of n repetitions. More generally, there may exist g homogeneous groups with the uth group consisting of n_u repetitions ($\Sigma n_u = n$) for which the ratings are $\pi_i^{(u)}$ ($i=1, 2, \ldots, t$; $u=1, 2, \ldots, g$). Within such a group the above procedures are applicable, and provide estimates of $\pi_i^{(u)}$ as well as a value $B_1^{(u)}$ of the B_1 statistic. If $\lambda^{(c)}$ is the likelihood ratio for the combined experiment, we now have in place of (4.4.3)

$$(4.4.4) \qquad -2\log\lambda^{(c)} = nt(t-1)\log 2 - 2B_1^{(c)}\log 10,$$

where

$$B_1^{(c)} = \sum_{u=1}^{g} B_1^{(u)}\log 10.$$

On the null hypothesis of complete randomness ($\pi_i^{(u)} = 1/t$, all i, u) $-2\log\lambda^{(c)}$ is distributed asymptotically as χ^2 with $g(t-1)$ DF. Whether (4.4.4) or (4.4.3) is to be used in a particular instance depends on one's prior information about possible group differences.

If the hypothesis of interest is not complete randomness but rather the absence of group differences ($\pi_i^{(u)} = \pi_i$, all u) then a large-sample test is obtained by referring

$$(4.4.5) \qquad -2\log(\lambda^{(c)}/\lambda) = -2(B_1^{(c)} - B_1)\log 10$$

to tables of χ^2 with $(g-1)(t-1)$ DF. Some caution in the use of this test seems advisable since the distribution of $-2\log(\lambda^{(c)}/\lambda)$ in finite samples depends on the nuisance parameters π_i.

The reader is referred to the papers by Bradley and Terry (1952) and Wilkinson (1957) for numerical examples and further discussion.

Treatment contrasts and factorials

Analyses of experiments where the objects are 2×2 factorial treatment combinations were already considered by Abelson and Bradley (1954). A simplified approach for general factorials was developed by Bradley and El-Helbawy (1976). Their starting point is to maximize the likelihood L of (4.3.1) when m ($< t$) orthogonal contrasts of the log π_i are zero. (Here it is convenient to replace $\Sigma\pi_i = 1$ by $\Sigma \log \pi_i = 0$.) The m contrasts can then be identified with factorial treatment combinations of interest and likelihood ratio tests made accordingly. We will not enter into details since Bradley (1984) has himself given a good summary, with an application to a 2^3 experiment. See also Littell and Boyett (1977) and El-Helbawy (1984).

The construction of 2^n-factorial paired-comparison designs that are optimal for detecting various factorial effects, single or interactions, is investigated by El-Helbawy and Bradley (1978) and El-Helbawy and Ahmed (1984). See also the concluding section of §4.6.

Mattenklott, Sehr, and Miescke (1982) use a factorial-type model for paired comparisons of multi-attribute stimuli by regarding each attribute as occurring at two or more levels.

4.5 Tests of the model

In a practical situation it will seldom be clear-cut which of the foregoing linear models—and of an infinity of others which might be considered—is actually appropriate. Fortunately there is a good deal of general evidence that preference scales constructed according to different laws tend to agree in essentials if *some* linear model applies. For an empirical study see, e.g., Jackson and Fleckenstein (1957) and Mosteller (1958). Instead of making a direct comparison of scales one may also investigate how well the estimated points on the scale, the v_i, serve to reproduce the observed preference proportions p_{ij}. Thus Thurstone (1927c) "reconstructs" estimates p'_{ij} of the preference probabilities π_{ij} by means of

$$p'_{ij} = H(v_i - v_j), \tag{4.5.1}$$

H being the law used in obtaining the v_i, and compares the p'_{ij} with the p_{ij}. This idea is useful and widely applicable, although it provides no formal criterion as to what constitutes a satisfactory reconstruction. Mosteller (1951c) has used it as the basis of an approximate χ^2 goodness of fit test for the Thurstone–Mosteller model. For a theoretical method of reconstruction see Noether (1960).

Mosteller's test of fit

The angular transformation (4.2.1) produces variates d_{ij} which for large n are approximately normal with variance $1/n$. If, correspondingly, we define

$$d'_{ij} = \sin^{-1}(2p'_{ij} - 1),$$

then

$$X^2 = n \sum_{i<j} (d_{ij} - d'_{ij})^{2(*)}$$

may be expected to have an approximate χ^2-distribution when the model under test is true. Since the d'_{ij} have not been obtained by a fully efficient method the degrees of freedom are somewhat uncertain, but we should not be seriously misled by taking them to be

$$\tfrac{1}{2}t(t-1)-(t-1)=\tfrac{1}{2}(t-1)(t-2).$$

The method is easily carried out not only in the normal case but also for other linear models. As Mosteller points out, there is, however, a tendency for the resulting fit to appear to be better than it is. A possible explanation lies in lack of constancy of the individual π_{ij} throughout the n replications of the experiment. It is well known (cf. §3.1) that this has the effect of reducing the variance of p_{ij} below its value for constant π_{ij}, and hence tends to reduce X^2. Another explanation is offered in §4.8.

Bradley's test of fit

A test of the Bradley–Terry model can be made directly from likelihood-ratio theory (Bradley, 1954a). Under

$$H_0 : \pi_{ij} = \pi_i/(\pi_i + \pi_j) \qquad (i, j = 1, 2, \ldots, t;\ i \neq j),$$

(*) This is identical with Mosteller's criterion. The angles d_{ij}, d'_{ij} are measured in radians.

the likelihood is given by (4.3.1), while under the alternative

$$H_1 : \pi_{ij} \neq \pi_i/(\pi_i + \pi_j) \quad \text{for some } i \text{ and } j$$

it is

$$C \prod_{i<j} \pi_{ij}^{\alpha_{ij}} \pi_{ji}^{\alpha_{ji}}.$$

Since p_{ij} is the ML estimate of π_{ij} on H_1, the likelihood ratio criterion for testing H_0 against H_1 is

$$\lambda = \frac{\prod_{i=1}^{t} p_i^{a_i}}{\prod_{i<j} (p_i + p_j)^n} \cdot \frac{1}{\prod_{i<j} p_{ij}^{\alpha_{ij}} p_{ji}^{\alpha_{ji}}},$$

or

$$-2 \log \lambda = 2\left[\sum_{i \neq j} \alpha_{ij} \log p_{ij} - \Sigma a_i \log p_i + n \sum_{i<j} \log(p_i + p_j)\right]$$

$$= 2\left[\sum_{i \neq j} \alpha_{ij} \log p_{ij} + B_1 \log 10\right],$$

with B_1 as in (4.4.2). H_0 is rejected at significance level α if $-2 \log \lambda$ exceeds the upper α significance point of χ^2 with $\frac{1}{2}(t-1)(t-2)$ DF. This test appears to share the practical difficulties of the Mosteller test in too readily accepting the model.

4.6 Incomplete contingency table approaches

Although the results of a paired-comparison experiment can be expressed concisely in a preference table, an alternative representation as an incomplete contingency table is sometimes useful. Table 4.2 displays this simple layout.

The purpose of Table 4.2, which provides the same information as a preference table, is to put paired-comparison experiments into the framework of contingency table methodology. However, all analyses so far proposed along these lines are based on the Bradley–Terry model.

Table 4.2 Incomplete contingency table representation of a paired-comparison experiment

Pair $i\ j$	No. of preferences 1	2	3	...	$t-1$	t	No. of comparisons n_{ij}
1 2	α_{12}	α_{21}	—		—	—	n_{12}
1 3	α_{13}	—	α_{31}		—	—	n_{13}
⋮							⋮
$t-1, t$	—	—	—		$\alpha_{t-1,t}$	$\alpha_{t,t-1}$	$n_{t-1,t}$
Total	a_1	a_2	a_3		a_{t-1}	a_t	Σa_i

We begin with a logistic modelling method due to Imrey, Johnson, and Koch (1976), an application of the large-sample weighted least-squares approach of Grizzle, Starmer, and Koch (1969). With $p_{ij} = \alpha_{ij}/n_{ij}$ $(i, j = 1, \ldots, t,\ i \neq j)$, define the logit transformation

$$
\begin{aligned}
u_{ij} &= \log(p_{ij}/p_{ji}) & 0 < p_{ij} < 1 \\
&= -\log(2n_{ij} - 1) & p_{ij} = 0 \qquad (4.6.1)\\
&= \log(2n_{ij} - 1) & p_{ij} = 1
\end{aligned}
$$

Here the $p_{ij} = 0$ or 1 cases incorporate the remedial action mentioned following (4.1.6). On the usual assumptions, all the u_{ij} for $i < j$ are clearly independent.

Under the Bradley–Terry model we have asymptotically, as each $n_{ij} \to \infty$, that

$$
\begin{aligned}
\mathscr{E}(u_{ij}) &\sim \log(\pi_{ij}/\pi_{ji}) \qquad (4.6.2)\\
&= \log(\pi_i/\pi_j) = \beta_{ij} \text{ (say)}.
\end{aligned}
$$

Also, expressing u_{ij} as a function of α_{ij}, one finds

$$\operatorname{var}(u_{ij}) \sim (n_{ij}\pi_{ij}\pi_{ji})^{-1}$$

for which a consistent estimator is $v_{ij} = (n_{ij}p_{ij}p_{ji})^{-1}$.

It is now possible to estimate the β_{ij} (and hence the π_i) by weighted least squares, leading to asymptotically optimal (BAN) estimates. To this end, note that $\beta_{ij} = \beta_{1j} - \beta_{1i}$ and $\beta_{ji} = -\beta_{ij}$. Let $\boldsymbol{u}' = (u_{12}, u_{13}, \ldots, u_{t-1,t})$, $\boldsymbol{\beta}' = (\beta_{12}, \beta_{13}, \ldots, \beta_{1t})$, and let $\boldsymbol{V}_u$ denote the estimated diagonal covariance matrix of $\boldsymbol{u}$ with elements v_{12}, $v_{13}, \ldots, v_{t-1,t}$. Also let $\boldsymbol{X}$ be the $\frac{1}{2}t(t-1)\times(t-1)$ design matrix

$$\boldsymbol{X}=\begin{bmatrix} 1 & 0 & 0 & & 0 & 0 \\ 0 & 1 & 0 & & 0 & 0 \\ & & & \cdots & & \\ 0 & 0 & 0 & & 0 & 1 \\ -1 & 1 & 0 & & 0 & 0 \\ -1 & 0 & 1 & & 0 & 0 \\ & & & \cdots & & \\ 0 & 0 & 0 & & -1 & 1 \end{bmatrix}$$

where the columns correspond to $\beta_{12}, \ldots, \beta_{1t}$ and the rows to the $\frac{1}{2}t(t-1)$ pairings. We now have asymptotically

$$\mathscr{E}(\boldsymbol{u}) \sim \boldsymbol{X\beta} \quad \text{and} \quad \operatorname{cov}(\boldsymbol{u}) \sim \boldsymbol{V}_u, \tag{4.6.3}$$

with weighted least-squares solution

$$\hat{\boldsymbol{\beta}}=(\boldsymbol{X}'\boldsymbol{V}_u^{-1}\boldsymbol{X})^{-1}\boldsymbol{X}'\boldsymbol{V}_u^{-1}\boldsymbol{u}.$$

Also the estimated covariance matrix of $\hat{\boldsymbol{\beta}}$ is

$$\boldsymbol{V}_{\hat{\beta}}=(\boldsymbol{X}'\boldsymbol{V}_u^{-1}\boldsymbol{X})^{-1}.$$

Estimates of the π_i follow from (4.6.2), namely,

$$\hat{\pi}_j=\hat{\pi}_1\,\mathrm{e}^{-\hat{\beta}_{1j}} \qquad \hat{\pi}_1=\left[1+\sum_{j=2}^{t}\mathrm{e}^{-\hat{\beta}_{1j}}\right]^{-1}. \tag{4.6.4}$$

A test statistic of the Bradley–Terry model now follows immediately from standard theory since

$$X^2=\boldsymbol{u}'\boldsymbol{V}_u^{-1}\boldsymbol{u}-\hat{\boldsymbol{\beta}}'(\boldsymbol{X}'\boldsymbol{V}_u^{-1}\boldsymbol{X})\hat{\boldsymbol{\beta}} \tag{4.6.5}$$

has an asymptotic χ^2 distribution with $\frac{1}{2}t(t-1)-(t-1)=\frac{1}{2}(t-1)(t-2)$ DF. The same methods have been applied by Zimmerman and Rahlfs (1976) and Beaver (1977a) to the Bradley–Terry model with ties (cf. §7.3), and by Beaver also to the B-T model allowing for order effects, and to triple comparison models.

See also Atkinson (1972) and Fienberg and Larntz (1976). The latter reference obtains the exact Bradley–Terry theory as a special case of contingency table analysis with either "quasi-symmetry" or "quasi-independence" but requires iterative methods (iterative proportional fitting).

Response surface fitting in factorials

Factorial paired-comparison experiments were touched on briefly in §4.4. Response surface fitting was first investigated by Springall (1973) in the case of the Rao–Kupper (1967) extension for ties (see §7.3) of the Bradley–Terry model. See also Berkum (1987). Beaver (1977b) applied the methods of this section to the same problem. Monte Carlo simulations were employed by Starks (1982) to test the adequacy of the large-sample theory. Specifically, Starks considers experiments where each factorial treatment combination is compared n times to a standard (and not to another treatment combination). He finds good agreement with asymptotic theory if $n \geqslant 20$; smaller values of n suffice if the preference probabilities are not too extreme. Of course, one might wish to consider other methods of experimentation, e.g., designs confounding higher-order interactions (Quenouille and John, 1971).

4.7 Multivariate paired comparisons

A versatile and relatively simple approach to multivariate paired comparisons when the univariate marginal distributions follow Bradley–Terry models is provided by methods of Imrey *et al.* (1976). We follow their exposition and consider an experiment, previously studied by Davidson and Bradley (1969), comparing three chocolate puddings on the basis of overall quality, taste, and colour. The experimental results are given in Table 4.3.

From the rows of the table we can generate three vectors such as

$$\boldsymbol{p}'_{12} = \frac{1}{20}(6, 2, 1, 1, 1, 1, 0, 8)$$

Table 4.3 Preferences on the basis of three characteristics

Pair* i	j	Pattern of preferences iii	iij	iji	ijj	jii	jij	jji	jjj	Total n_{ij}
1	2	6	2	1	1	1	1	0	8	20
1	3	3	2	1	1	1	2	1	11	22
2	3	12	2	0	0	0	1	0	8	23

*1 = overall quality, 2 = taste, 3 = colour

which has a multinomial distribution with

$$\mathscr{E}(\boldsymbol{p}'_{12}) = \boldsymbol{\pi}_{12}, \text{ say,}$$

and

$$\operatorname{cov}(\boldsymbol{p}'_{12}) = \frac{1}{20}(\boldsymbol{D}_{\pi_{12}} - \boldsymbol{\pi}_{12}\boldsymbol{\pi}'_{12}) = \boldsymbol{V}_{\pi_{12}},$$

where $\boldsymbol{D}$ is a diagonal matrix with the components of $\boldsymbol{\pi}_{12}$ as elements. Also an asymptotically unbiased estimator of $\boldsymbol{V}_{\pi_{12}}$ is given by

$$\boldsymbol{V}_{p_{12}} = (1/20)(\boldsymbol{D}_{p_{12}} - \boldsymbol{p}_{12}\boldsymbol{p}'_{12}).$$

Thus, $\boldsymbol{p}' = (\boldsymbol{p}'_{12}, \boldsymbol{p}'_{13}, \boldsymbol{p}'_{23})$ has expectation $\boldsymbol{\pi}' = (\boldsymbol{\pi}'_{12}, \boldsymbol{\pi}'_{13}, \boldsymbol{\pi}'_{23})$ and covariance matrix $\boldsymbol{V}_\pi$ which may be consistently estimated by

$$\boldsymbol{V}_p = \begin{pmatrix} \boldsymbol{V}_{p_{12}} & \boldsymbol{0} & \boldsymbol{0} \\ \boldsymbol{0} & \boldsymbol{V}_{p_{13}} & \boldsymbol{0} \\ \boldsymbol{0} & \boldsymbol{0} & \boldsymbol{V}_{p_{23}} \end{pmatrix}$$

Now let $\boldsymbol{K}_{9\times 18} = [1 \quad -1] \otimes I_9$, where the Kronecker product symbol $\otimes$ here means that each element of I_9 is to be multiplied by $[1 \quad -1]$. Also let

$$\boldsymbol{A}^* = \begin{bmatrix} 1 & 1 & 1 & 1 & 0 & 0 & 0 & 0 \\ 0 & 0 & 0 & 0 & 1 & 1 & 1 & 1 \\ 1 & 1 & 0 & 0 & 1 & 1 & 0 & 0 \\ 0 & 0 & 1 & 1 & 0 & 0 & 1 & 1 \\ 1 & 0 & 1 & 0 & 1 & 0 & 1 & 0 \\ 0 & 1 & 0 & 1 & 0 & 1 & 0 & 1 \end{bmatrix}, \quad \underset{18\times 24}{\boldsymbol{A}} = \boldsymbol{A}^* \otimes \boldsymbol{I}_3,$$

and

$$\boldsymbol{u} = \boldsymbol{K}\,\mathbf{log}(\boldsymbol{Ap}).$$

Here $\mathbf{log}(\boldsymbol{Ap})$ is the vector whose first of 18 elements is $\log p_{12,1}$, where $p_{12,1}$ is the proportion $(6+2+1+1)/20$ of preferences for the first over the second pudding on the basis of criterion 1. Correspondingly, the first of 9 elements of $\boldsymbol{u}$ is

$$u_{12,1} = \log(p_{12,1}/p_{21,1}),$$

the logit transform of (4.6.1). We see that

$$\boldsymbol{u}' = (u_{12,1}, u_{12,2}, u_{12,3}, \ldots, u_{23,1}, u_{23,2}, u_{23,3}).$$

Since asymptotically, for positive variates X_1 X_2,

$$\operatorname{cov}(\log X_1, \log X_2) = \operatorname{cov}(X_1, X_2)/\mathscr{E}X_1\ \mathscr{E}X_2,$$

it follows that the covariance matrix of $\boldsymbol{u}$ may be estimated consistently by

$$\boldsymbol{V}_u = \boldsymbol{K}\boldsymbol{D}_{\boldsymbol{Ap}}^{-1}(\boldsymbol{A}\boldsymbol{V}_p\boldsymbol{A}')\boldsymbol{D}_{\boldsymbol{Ap}}^{-1}\boldsymbol{K}'.$$

We can now fit the marginal Bradley–Terry models

$$\pi_{ij,\alpha} = \frac{\pi_{i,\alpha}}{\pi_{i,\alpha} + \pi_{j,\alpha}} \qquad i, j = 1, 2, 3,\ i \neq j,\ \alpha = 1, 2, 3$$

by applying weighted least squares to (4.6.3) with

$$\boldsymbol{\beta}' = (\beta_{12,1}, \beta_{13,1}, \beta_{12,2}, \beta_{13,2}, \beta_{12,3}, \beta_{13,3}),$$

where

$$\beta_{ij,\alpha} = \log(\pi_{i,\alpha}/\pi_{j,\alpha}),$$

and with

$$\boldsymbol{X} = \begin{bmatrix} 1 & 0 & 0 & 0 & 0 & 0 \\ 0 & 0 & 1 & 0 & 0 & 0 \\ 0 & 0 & 0 & 0 & 1 & 0 \\ 0 & 1 & 0 & 0 & 0 & 0 \\ 0 & 0 & 0 & 1 & 0 & 0 \\ 0 & 0 & 0 & 0 & 0 & 1 \\ -1 & 1 & 0 & 0 & 0 & 0 \\ 0 & 0 & -1 & 1 & 0 & 0 \\ 0 & 0 & 0 & 0 & -1 & 1 \end{bmatrix}$$

From the resulting estimates $\hat{\beta}_{ij,\alpha}$ one can solve for the $\hat{\pi}_{j,\alpha}$ as in (4.6.4). The goodness-of-fit X^2 of (4.6.5), summed over the three characteristics, is 3.05, with 3 DF, indicating that the Bradley–Terry model provides a good overall fit to the univariate marginal distributions. For more detailed results and comparisons with other estimation methods see Imrey *et al.* (1976).

The foregoing approach is appropriate when primary interest lies in the parameters of the marginal distributions. Although the analysis recognizes the multivariate nature of the data, there are no parameters of association between the various characteristics. A multivariate generalization of the Bradley–Terry model, specifically incorporating parameters of association between pairs of characteristics, is proposed and investigated in some detail by Davidson and Bradley (1969, 1970). Recall also the nonparametric procedure of Sen and David (1968) presented in §2.5.

4.8 Bock's three-component model

The linear models so far discussed do not make any allowance for possible differences between judges. Bock (1958) has proposed that the observed merits of object A_i in its comparison with A_j by judge γ may be represented by

$$(4.8.1) \qquad y_{i\gamma(j)} = V_i + w_{i\gamma} + z_{i\gamma(j)} \qquad (i, j = 1, 2, \ldots, t;\ i \neq j;\ \gamma = 1, 2, \ldots, n),$$

where the brackets around j indicate that this subscript serves merely as a label. Small letters denote random variables, the $z_{i\gamma(j)}$ being mutually independent $N(0, \delta^2)$ variates. The special feature of this model lies in the $w_{i\gamma}$ which are components peculiar to specific objects and judges, independent over the sample of judges. For the same judge they are correlated, $w_{i\gamma}$ and $w_{j\gamma}$ having a bivariate normal $N(0, 0, \zeta^2, \zeta^2, \rho)$ distribution. This assumption is equivalent (in the usual case $\rho \geqslant 0$) to representing $w_{i\gamma}$ as the sum of a normally distributed judge effect plus a normally distributed interaction effect. The "mixed" nature of the model reflects specific interest in the objects, but interest in the judges only as a sample from an infinite population of judges. That the model is linear follows if we note that $d_{ij\gamma} = y_{i\gamma(j)} - y_{j\gamma(i)}$ is normally distributed with mean $V_i - V_j$ and variance $2\zeta^2(1-\rho) + 2\delta^2$. Taking this variance to be unity we obtain

$$\pi_{ij} = \Pr(d_{ij\gamma} > 0)$$
$$= H(V_i - V_j),$$

with H the unit normal cdf as for the Thurstone–Mosteller model. It may now appear that the two models are the same after standardization. However, in the present case comparisons sharing a common object and made by the same judge are not independent since

$$\operatorname{cov}(d_{ij\gamma}, d_{ik\gamma}) = \zeta^2(1-\rho) \qquad (i \neq j \neq k).$$

In terms of the characteristic random variables $x_{ij\gamma}$ of §1.3 we have

$$\Pr(x_{ij\gamma}=1)=\pi_{ij}, \quad \Pr(x_{ij\gamma}=0)=\pi_{ji},$$

and $$\Pr(x_{ij\gamma}=1, x_{ik\gamma}=1)=\pi_{ij,ik} \quad \text{(say)}.$$

Since $$p_{ij}=\sum_{\gamma} x_{ij\gamma}/n$$

it follows that

$$\mathscr{E}(p_{ij})=\pi_{ij}, \quad \operatorname{var}(p_{ij})=\pi_{ij}\pi_{ji}/n,$$

and $$\operatorname{cov}(p_{ij}, p_{ik})=(\pi_{ij,ik}-\pi_{ij}\pi_{ik})/n.$$

for the reasons stated in §4.2 Bock applies the angular transformation (4.2.1) to the p_{ij}. While this transformation approximately stabilizes the variance for large n it will, like all such transformations, leave the correlation essentially unchanged; that is,

$$\operatorname{corr}(d_{ij}, d_{ik}) \doteqdot \operatorname{corr}(p_{ij}, p_{ik})=\frac{\pi_{ij,ik}-\pi_{ij}\pi_{ik}}{\sqrt{(\pi_{ij}\pi_{ji}\pi_{ik}\pi_{ki})}}.$$

In order to obtain a workable solution for his three-component model Bock assumes that this correlation (ρ', say) is the same for all i, j, k—an assumption which, as he says, is not good in general but which is reasonable when all π_{ij} are close to $\frac{1}{2}$. It is just for such closeness to the null situation that even for large n a sensitive test for the equality of the objects is required.

Subject to the various approximations involved, we can now express d_{ij} in the form

$$d_{ij}=\alpha_i+\beta_j+e_{ij} \qquad (i, j=1, 2, \ldots, t), \tag{4.8.2}$$

where α_i is the merit of A_i on the transformed scale, and $\beta_j=-\alpha_j$. It is convenient to define $d_{ii}=0$. The random error terms e_{ij} have therefore the following properties:

$$e_{ij}=-e_{ji}, \quad \mathscr{E}(e_{ij})=0, \quad \mathscr{E}(e_{ij}^2)=\sigma^2=1/n, \tag{4.8.3}$$

and for i, j, k, l unequal

$$\mathscr{E}(e_{ij}e_{ik})=\rho'\sigma^2, \quad \mathscr{E}(e_{ij}e_{kl})=0. \tag{4.8.4}$$

If
$$d_i = \sum_{j=1}^{t} d_{ij}$$

the sum of squares between rows of the matrix (d_{ij}) is

$$S_\alpha = \Sigma d_i^2 / t$$

and equals the sum of squares between columns S_β. From (4.8.2) we have

$$S_\alpha = \frac{1}{t} \sum_i \left(t\alpha_i + \sum_j e_{ij} \right)^2,$$

so that, in view of (4.8.3) and (4.8.4),

$$\mathscr{E}(S_a) = t\Sigma \alpha_i^2 + (t-1)\sigma^2 + (t-1)(t-2)\rho'\sigma^2.$$

Also the residual sum of squares

$$S_R = \Sigma d_{ij}^2 - S_\alpha - S_\beta$$

has expected value

$$(t-1)(t-2)\sigma^2 - 2(t-1)(t-2)\rho'\sigma^2.$$

Equivalently, these results may be set out as an analysis of variance as follows:

Source of variation	Degrees of freedom	Sums of squares	Expected sum of squares
Between objects	$t-1$	$S_\alpha = \frac{1}{t}\Sigma d_i^2$	$(t-1)[1+(t-2)\rho']/n + t\Sigma a_i^2$
Residual	$\frac{1}{2}(t-1)(t-2)$	$\frac{1}{2}S_R = \frac{1}{2}S_T - S_\alpha$	$(t-1)(t-2)(1-2\rho')/(2n)$
Total	$\frac{1}{2}t(t-1)$	$\frac{1}{2}S_T = \sum_{i<j} d_{ij}^2$	

Except for the approximation of the normal law by the angular, Mosteller's test of fit is equivalent to treating nS_R as a χ^2 with $\frac{1}{2}(t-1)(t-2)$ DF. But for $\rho' > 0$ we see that S_R has too small an expectation. Bock's model, therefore, is able to account for the aberrant behaviour of Mosteller's test.

It may also be noted from the table that if equality of the objects is tested by regarding $n\Sigma d_i^2/t$ as a χ^2 with $t-1$ DF, we may be led to reject too easily for $\rho' > 0$.

The design and analysis of incomplete paired-comparison designs suitable for estimating the parameters of this model are treated by Linhart (1966); see also Bock and Jones (1968, p. 179).

Notes

(1) Our treatment of the important Thurstone–Mosteller model and of the other "Cases" (cf. Exercise 1.5) distinguished by Thurstone (1927a) has been kept brief in view of the summary accounts by Torgerson (1958) and Bock and Jones (1968). Research relating to the Bradley–Terry model has been reviewed by Bradley (1976, 1984). Our discussion of this model continues in §7.3 (treatment of ties) and §7.4 (within-pair order effects).

(2) Maximum likelihood estimation for Luce's model (1.4.1), which includes the Bradley–Terry model, is handled by iterative proportional fitting in van Putten (1982).

(3) Winsberg and Ramsay (1981) have proposed the use of integrated *B*-splines for the analysis of data satisfying an unspecified linear model.

EXERCISES

4.1 Analyze the data on p. 116 by assuming a Bradley–Terry model. Also test the goodness of fit of your model.

[Answers: $p_1 = 0{\cdot}049$, $p_2 = 0{\cdot}248$, $p_3 = 0{\cdot}181$, $p_4 = 0{\cdot}522$; $-2\log\lambda = 30{\cdot}45$ (4.4.3); $-2\log\lambda = 4{\cdot}24$ (4.5.2)]

4.2 Analyze the same data by the methods of §4.6.

[Answers: $\hat{\pi}_1 = 0{\cdot}053$, $\hat{\pi}_2 = 0{\cdot}253$, $\hat{\pi}_3 = 0{\cdot}187$, $\hat{\pi}_4 = 0{\cdot}521$; $X^2 = 3{\cdot}91$]

5 DESIGNS

5.1 Completely balanced designs

In the language of the design of experiments, what we have called a balanced paired-comparison experiment is known as a "balanced incomplete block" (BIB) design, the block size being two. The problem of design is, of course, the same whether we have for each block an expression of preference, as in a paired comparison, or two separate values as in the standard experimental situation. There is also a close correspondence between a balanced paired-comparison experiment and a Round Robin tournament as used in many sports, where each of t players (or teams) plays every other player once or several times. The design of such a completely balanced experiment presents no difficulties, but a particularly simple approach given by Kraitchik (1953) may be of interest. It is convenient to label the players $0, 1, \ldots, t-1$. His procedure may be illustrated for $t=7$. Write the numbers 0–6 in sequence in seven rows of four $(=\frac{1}{2}(t+1))$. This yields the four columns of figures in heavy type. Then proceed backwards in the

(7)	**0**		6	**1**	5	**2**	4	**3**
	4	(7)	3	**5**	2	**6**	1	**0**
(7)	**1**		0	**2**	6	**3**	5	**4**
	5	(7)	4	**6**	3	**0**	2	**1**
(7)	**2**		1	**3**	0	**4**	6	**5**
	6	(7)	5	**0**	4	**1**	3	**2**
(7)	**3**		2	**4**	1	**5**	0	**6**

same way, starting with the bottom right-hand corner and inserting the numbers in three columns as shown by the figures in ordinary type. The result is an arrangement for a Round Robin tournament of 7 players requiring 7 rounds. The player named in the first column has a bye. It will be noted that every player appears first in 3 out of 6 games, corresponding to playing 3 games at home or with the white pieces, etc. From the point of view of a paired-comparison experiment we have, on ignoring the first column, a design in which any effect due to order of presentation is balanced out, and which incorporates other safeguards. The process works for any odd value of t. If t is even, say $t=8$, we simply pair off the eighth player with the player having the bye, writing 7 alternately to the left and to the right of the first column, as shown. The effect of order of presentation is now balanced-out only approximately; a complete balancing-out would be achieved by repeating the whole experiment with the order within each pair reversed and the rows re-randomized. For another simple approach see Freund (1956).

A method that maintains the greatest possible separation between pairs having an object in common has been given by Ross (1934). For t odd write out the pairs as follows:

$$
\begin{array}{cccccc}
\left\{\begin{array}{c} 01 \\ 0, \tfrac{1}{2}(t+1) \end{array}\right. & \begin{array}{c} t-1, 2 \\ 12 \end{array} & \begin{array}{c} t-2, 3 \\ t-1, 3 \end{array} & \begin{array}{c} t-3, 4 \\ t-2, 4 \end{array} & \begin{array}{c} \ldots \\ \ldots \end{array} & \begin{array}{c} \tfrac{1}{2}(t+3), \tfrac{1}{2}(t-1) \\ \tfrac{1}{2}(t+3), \tfrac{1}{2}(t+1) \end{array} \\
\left\{\begin{array}{c} 02 \\ 0, \tfrac{1}{2}(t+3) \end{array}\right. & \begin{array}{c} 13 \\ 23 \end{array} & \begin{array}{c} t-1, 4 \\ 14 \end{array} & \begin{array}{c} t-2, 5 \\ t-1, 5 \end{array} & \begin{array}{c} \ldots \\ \ldots \end{array} & \begin{array}{c} \tfrac{1}{2}(t+5), \tfrac{1}{2}(t-3) \\ \tfrac{1}{2}(t+5), \tfrac{1}{2}(t+3) \end{array} \\
 & & \cdots & & & \\
\left\{\begin{array}{c} 0, \tfrac{1}{2}(t-1) \\ 0, t-1 \end{array}\right. & \begin{array}{c} \tfrac{1}{2}(t-3), \tfrac{1}{2}(t+1) \\ \tfrac{1}{2}(t-1), \tfrac{1}{2}(t+1) \end{array} & \begin{array}{c} \tfrac{1}{2}(t-5), \tfrac{1}{2}(t+3) \\ \tfrac{1}{2}(t-3), \tfrac{1}{2}(t+3) \end{array} & \begin{array}{c} \tfrac{1}{2}(t-7), \tfrac{1}{2}(t+5) \\ \tfrac{1}{2}(t-5), \tfrac{1}{2}(t+5) \end{array} & \begin{array}{c} \ldots \\ \ldots \end{array} & \begin{array}{c} 1, t-2 \\ 1, t-1 \end{array}
\end{array}
$$

For $t=7$ we have, after balancing the order of presentation: 01, 62, 53, 40, 21, 36, 45, 02, 13, 64, 50, 32, 41, 56, 03, 24, 15, 60, 34, 25, 16. The case $t=6$ can be obtained from this (without complete balance) by omitting all pairs containing a 6. Ross has tabulated balanced orders for odd t up to 17. The structure of such greatest possible separation designs has been thoroughly examined by Simmons and Davis (1975).

Other balanced arrangements which may be of value for special purposes can be found in Haselgrove and Leech (1977), Archbold and Johnson (1958), and Chakravarti (1976).

A feature of Kraitchik's design that is unsatisfactory in certain applications may be seen by noting that player 1 meets in rows 2 to 7 players who have just met 7, 4, 0, 0, 0, 0, respectively. Thus if, for example, 0 is a strong player, then 1 has the repeated advantage of meeting opponents immediately after they have had a tough match. Player 1 may be said to receive a *carry-over effect* from 7, 4, and 0 in rows 2, 3, 4–7, respectively. Russell (1980) has studied designs that balance such effects either exactly or approximately. Exact balance is possible when t is a power of 2. Thus for $t=8$ he obtains by Galois field theory

0 3	1 4	2 7	5 6
0 4	1 3	2 5	6 7
0 5	1 7	2 4	3 6
0 6	1 2	3 5	4 7
0 7	1 5	2 3	4 6
0 1	2 6	3 4	5 7
0 2	1 6	3 7	4 5

This design can easily be balanced approximately for order effects.

The design of doubles tournaments has also been considered by Kraitchik and by Scheid (1960). Bose and Cameron (1965) gave bridge tournament designs for up to 50 players. These plans, derived from BIB designs, arrange the t players into b blocks of size 4, each block consisting of two half-blocks of size 2. Two players appear in the same half-block λ_1 times and in opposite half-blocks of the same block λ_2 times. Gilbert (1961) and Yalavigi (1967) have studied mixed-doubles tournaments. Doubles tournaments are of some interest in paired-comparison work when we wish to compare one pair of objects with another pair. For example, in establishing a colour-scale observers are asked to compare *differences* (is the difference between A_1 and A_2 greater than that between A_3 and A_4?). The mixed-doubles situation arises when the two objects making up one side of the comparison are of two different types (do you prefer food A_1 with drink A_2 to food A_3 with drink A_4?). However, the analogy is imperfect since a player can be on one side only, whereas with objects we may also wish to run comparisons of the type (A_1, A_2) *v.* (A_1, A_3).

5.2 Incomplete designs

A practical difficulty in the use of paired-comparison experiments is the large number of comparisons required when the objects are numerous. We may sympathize with the foreman whom McCormick *et al.* (1952) presented with a deck of 1,225 cards, each of which contained a different pairing of 50 employees working under him and each of which required an expression of preference. The authors investigated empirically the effect of omitting various fractions of the comparisons, using for this purpose incomplete designs of a cyclic type, as illustrated in Fig. 5.1. The pattern is clear and corresponds to omitting $\frac{3}{7}$ of the comparisons. In this example of personnel-rating the reader may wonder about the appropriateness of paired comparisons. Certainly the individual judgments are not completely independent if the foreman has a good memory, but with so many employees this effect should not be too important under proper randomization of the order of the pairs. Why not a straightforward ranking? With the large number of subjects involved this is not practicable, and in any case would give us no clue as to the foreman's judging ability. However, it would be possible to break down the comparisons into groups of five, say, instead of two. Again a selection of the $\binom{50}{5}$ possible blocks of five has to be made leading to BIB designs. It is not proper to replace the ranking within a block by the ten (consistent) paired comparisons

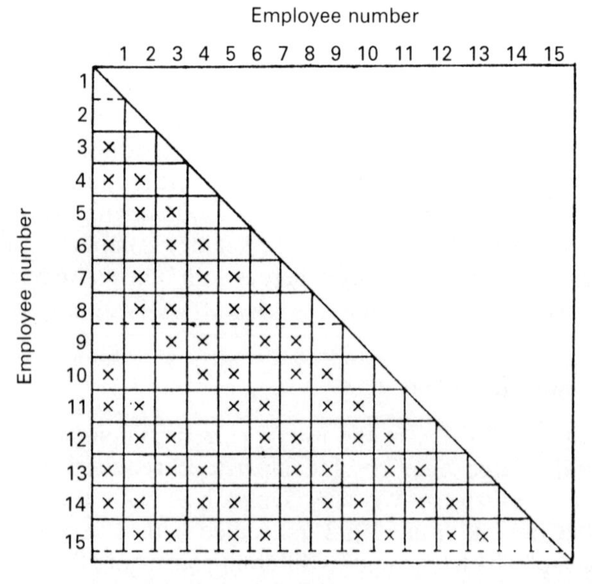

Fig. 5.1 **Illustration of a cyclic design; x denotes a pairing**

deducible from them, as has sometimes been suggested. An appropriate method of analysis is given by Durbin (1951), who proposes the use of Youden square (i.e., incomplete Latin square) designs since these compact BIB designs eliminate the effect of order of presentation within a block. See also Gulliksen and Tucker (1961).

A more clear-cut case for incomplete paired-comparison designs arises when the block size is of necessity restricted to two. The problem of what constitutes a satisfactory subset of the comparisons has been considered by Kendall (1955) who lays down the following two minimum requirements:

(a) every object should appear equally often;
(b) the preferences should not be divisible in the sense that we can split the objects into two sets and no comparison is made between any object in one and any object in the other.

5.3 Cyclic designs

For $t=7$, say, the 21 distinct pairs of objects may be set out as follows:

	{1}:	01	12	23	34	45	56	60
(5.3.1)	{2}:	02	13	24	35	46	50	61
	{3}:	03	14	25	36	40	51	62

From any pair its neighbour to the right may be obtained by increasing each object label by 1 and reducing modulo 7. Performing this operation on the last column of (5.3.1) takes us back to the first column. Each row has been arranged to start with the pair of lowest numerical value and is designated by the non-zero element in the initial pair placed in braces. Note that the order effects in each row (or *cyclic set*) are balanced out, giving two-way elimination of heterogeneity.

For t even, one always gets a "half-set" in addition to $\frac{1}{2}(t-1)$ full sets; e.g., for $t=6$ we have

{1}:	01	12	23	34	45	50
{2}:	02	13	24	35	40	51
{3}:	03	14	25			

Note that $\{2\}$ is not connected and separates into two subsets 02 24 40 and 13 35 51, the reason being that 2 is a factor of 6. In $\{3\}$, order effects are, of course, not balanced out and moreover $\{3\}$ is completely disconnected. Nevertheless, such disconnected sets may be useful as part of a connected design.

Some generalizations are apparent. When t is odd, the $\binom{t}{2}$ distinct pairs may be divided into $\frac{1}{2}(t-1)$ sets of t; for t even into $\frac{1}{2}(t-1)$ sets of t and one set of $\frac{1}{2}t$. If we ignore order effects, we have $\{x\} \sim \{t-x\}$, $x=1, 2,\ldots, m$, where $m=\frac{1}{2}(t-1)$ or $\frac{1}{2}t$ according as t is odd or even. The leading set $\{1\}$ will always be connected; so will $\{x\}$ provided x and t are relative primes. If x and t have greatest common divisor (g.c.d.) f, then $\{x\}$ separates into f subsets. A combination of sets $\{x_1\}, \{x_2\},\ldots$ gives a connected design if $t, x_1, x_2,\ldots$ have gcd unity.

When t is prime (>2), the sets $\{x\}$ ($x=1, 2,\ldots, m$) can be shown to be equivalent (David, 1963) in the sense that any set can be transformed into any other by suitable relabelling of the objects. For example, in (5.3.1) the renumbering permutation $R(7, 2)$ that multiplies the label of each of the 7 objects by $2 \pmod 7$ changes $\{1\}$ to

02 24 46 61 13 35 50

which is set $\{2\}$. Applying $R(7, 2)$ again, we obtain

04 41 15 52 26 63 30

which might be designated $\{4\}$ in the first instance but is equivalent to $\{3\}$. In general, the permutations $R(t, 1), R(t, 2),\ldots, R(t, m)$ form a group that is isomorphic with the multiplicative group of the integers, G: $1, 2,\ldots, m$, when products of these are reduced not only mod t but also for equivalences E of the type $i \equiv t-i$.

The group G may also be expressed as

$$G: 1, g, g^2,\ldots, g^{\frac{1}{2}(t-3)} \quad (\text{mod } t; E),$$

where g is a primitive root (mod t; E) of t; i.e., g is an integer such that $g^y \not\equiv 1$ (mod t; E) for $y=1, 2,\ldots, m-1$ but $g^m \equiv 1$ (mod t; E). We may now imagine the sets $\{x\}$ to be arranged around a circle in the order shown in Fig. 5.2. Each application of the permutation $R(t, g)$ then produces a single clockwise turn.

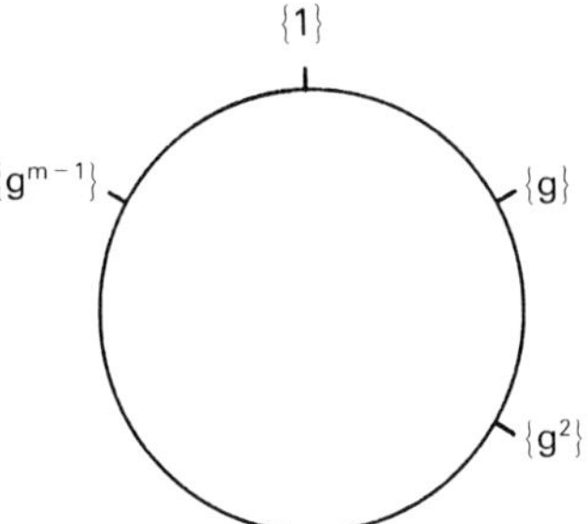

Fig. 5.2 Cyclic sets arranged on a circle

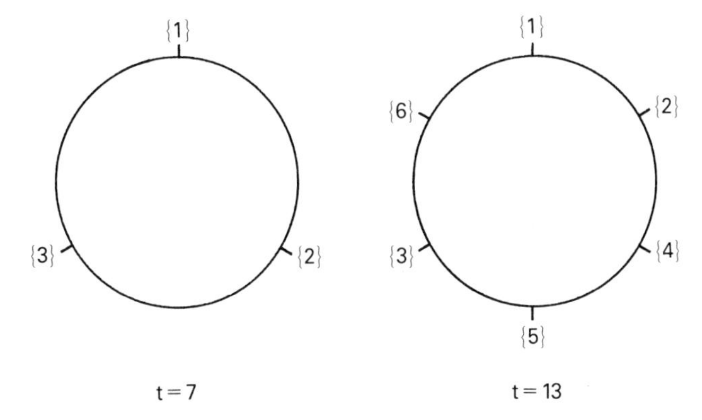

Fig. 5.3 Cyclic sets for $t=7$, 13

A design of size (t, r) (t objects replicated r times) is obtained by selecting $s=\frac{1}{2}r$ of the m points along the circumference. The situation is illustrated in Fig. 5.3 for $t=7$ and $t=13$. If $r=4$, two sets must be chosen, which can be done in only one distinct way for $t=7$ and in three distinct ways for $t=13$. In the latter case the

$\binom{6}{2}$ pairs of sets fall into the three classes:

$$\{1, 2\} \sim \{2, 4\} \sim \{4, 5\} \sim \{5, 3\} \sim \{3, 6\} \sim \{6, 1\}$$

$$\{1, 4\} \sim \{2, 5\} \sim \{4, 3\} \sim \{5, 6\} \sim \{3, 1\} \sim \{6, 2\}$$

$$\{1, 5\} \sim \{2, 3\} \sim \{4, 6\}$$

Similar methods of construction are possible when t is not prime.

For t prime the number of non-isomorphic designs of size (t, r) is equal to the number, $N(s, s')$, of distinct arrangements of s white and $s' = m - s$ black beads on a necklace (which may not be turned over). The formula, commonly attributed to Jablonski (1892) (see F. N. David and Barton, 1962), is

$$N(s, s') = \frac{1}{m} \Sigma \phi(d) \frac{(m/d)!}{(s/d)!(s'/d)!},$$

where the summation is over all positive integers d that are divisors of both s and s', and where $\phi(d)$ is Euler's function, the number of positive integers less than and prime relative to d, with $\phi(1) = 1$.

The more complicated problems of enumeration when t is not prime are treated in David (1972).

It will be clear that the class of cyclic designs is quite large. In fact, cyclic designs form a cross-section of partially balanced incomplete block designs with various numbers of associate classes. A great advantage of cyclic designs is that no plan of the experimental layout is needed since the initial block or blocks suffice. Table 5.1 lists the most efficient cyclic paired-comparison designs for $t \leqslant 14$. John, Wolock, and David (1972) cover $t \leqslant 30$. In this reference these designs constitute the case $k = 2$ of cyclic designs with block size k. See also John (1981) for a listing of efficient (but not necessarily maximally efficient) designs arranged to be robust in the sense that if the experiment is terminated after an even number of replications, the design used will have been close to maximal efficiency within the class of cyclic design.

Efficiency

By efficiency, E, in the preceding paragraph we mean the standard measure for an incomplete block design, namely the mean variance of treatment differences divided into $2\sigma^2/r$, the corresponding variance for a randomized block design with r replications. Figures of E obtained in this way are also relevant for cyclic preference experiments on t objects replicated r times (McKeon, 1960). It can be shown (e.g., John *et al.*, 1972) that

Table 5.1 Most efficient cyclic designs for $t \leqslant 14$

t	r		t	r	
5	2	{1}*	12	2	{1}
6	2	{1}		3	{1, 6}
	3	{1, 3}*		4	{2, 3}
	4	{1, 2}*		5	{1, 4, 6}
7	2	{1}		6	{1, 3, 5}*
	4	{1, 2}		7	{1, 3, 5, 6}
8	2	{1}		8	{1, 2, 4, 5}*
	3	{1, 4}		9	{1, 2, 3, 5, 6}*
	4	{1, 3}*		10	{1, 2, 3, 4, 5}*
	5	{1, 2, 4}	13	2	{1}
	6	{1, 2, 3}*		4	{1, 5}
9	2	{1}		6	{1, 2, 5}
	4	{1, 3}		8	{1, 2, 3, 6}
	6	{1, 2, 4}*		10	{1, 2, 3, 4, 5}
10	2	{1}	14	2	{1}
	3	{1, 5}		3	{1, 7}
	4	{1, 4}		4	{1, 4}
	5	{1, 3, 5}*		5	{1, 4, 7}
	6	{1, 2, 3}		6	{1, 3, 5}
	7	{1, 2, 3, 5}		7	{1, 2, 4, 7}*
	8	{1, 2, 3, 4}*		8	{1, 2, 5, 6}
11	2	{1}		9	{1, 2, 3, 6, 7}
	4	{1, 3}		10	{1, 2, 3, 4, 5}
	6	{1, 2, 4}		11	{1, 2, 3, 4, 5, 7}
	8	{1, 2, 3, 4}		12	{1, 2, 3, 4, 5, 6}

*Design is also group-divisible PBIB(2).

$$E = \frac{t-1}{rtc_{11}}, \quad \text{where} \quad c_{11} = \frac{1}{t}\sum_{h=2}^{t}\frac{1}{\Lambda_h} \tag{5.3.2}$$

and

$$\Lambda_j = a_1 + \sum_{h=2}^{t} a_h \cos(h-1)(j-1)\theta \quad j = 2, 3, \ldots, t$$

Here $\theta = 2\pi/t$ and

$$a_1 = r(k-1)/k,\ a_2 = a_t = -\lambda_{12}/k,\ a_3 = a_{t-1} = -\lambda_{13}/k, \ldots$$

$\ldots a_{(t+1)/2} = a_{(t+3)/2} = -\lambda_{1,(t+1)/2}/k$, if t is odd, or

$\ldots a_{(t+2)/2} = -\lambda_{1,(t+2)}/k$, if t is even, where λ_{ij} is the number of pairs in which both objects i and j occur.

The designs listed in John *et al.* (1972) are the most efficient cyclic designs of their size. Also given for each design are the efficiency factors $E(1, j)$, $j = 2, 3, \ldots, [\frac{1}{2}t] + 1$, for pairwise comparison of objects A_1 and A_j. (In this reference the objects are labelled from 1 to t.)

It may be noted that the same design problems arise in partial diallel cross experiments when it is not possible to make all $\frac{1}{2}t(t-1)$ single crosses among t inbred lines. A recent paper by Arya (1983) gives references to earlier work beginning with Kempthorne (1953).

A more general study of the efficiency of incomplete, including randomly incomplete, paired-comparison designs is made by Spence and Domoney (1974). See also Paterson (1983) and Paterson and Wild (1986).

Resolvable paired-comparison designs

An incomplete block design is said to be *resolvable* (Bose, 1942) if the blocks can be grouped so that each group contains every object exactly once, thus forming a complete replicate. An experiment laid out as a resolvable design can be performed in natural stages, a replicate at a time. When individual paired comparisons are laborious, a replicate may be a suitable unit to assign to a particular judge. Resolvability is at a premium whenever, as in a Round Robin tournament, the same t objects, and not merely replicas of them, have to be used throughout the experiment.

Many cyclic paired-comparison designs with t even are easily set out as resolvable designs. For example, if $t = 8$, $r = 5$, the design {1, 2, 4} may be written as

{1}:	01	23	45	67
	12	34	56	70
{2}:	02	46	13	57
	24	60	35	71
{4}:	04	15	26	37

Note that the order of presentation is as close to balance as is possible.

Quite generally we see that for t even and x relatively prime to t, the connected set $\{x\}$ can be split into two rounds. If t and x $(1 < x < \frac{1}{2}t)$ have greatest common divisor f, the set $\{x\}$ separates into f connected parts and is resolvable, provided t/f is even. The half-set $\{\frac{1}{2}t\}$ is always a replicate. Resolvable cyclic designs for $t \leqslant 14$ and their efficiencies are listed in David (1967).

For $r \leqslant \frac{1}{2}t$ Williams (1976) has proposed a class of resolvable paired-comparison designs ($\boldsymbol{\alpha}$-designs) which are constructed from cyclic designs for $s = \frac{1}{2}t$ objects and block size r but which are usually not themselves cyclic. For certain parameter combinations his designs are more efficient than the corresponding cyclic designs (but not, as claimed, for $t = 8$, $r = 4$ and $t = 12$, $r = 6$).

Resolvable paired-comparison designs for 2^n factorials are given by Quenouille and John (1971). Designs, not necessarily resolvable, for mixed factorials are treated by Lewis and Tuck (1985) and Gupta (1987). See also the final paragraph of §4.4.

Cyclic and related designs for general block size are treated in detail in a recent monograph by John (1986).

5.4 Partially balanced incomplete block designs

The relation of cyclic designs to partially balanced incomplete block (PBIB) designs with two associate classes is of some interest. For the present case, where each block consists of a pair of objects, PBIB(2) designs are defined as follows:

(a) There are t objects and b blocks.
(b) Each of the t objects occurs r times (so that $tr = 2b$).
(c) Every pair of objects occurs either λ_1 or λ_2 times (and the objects are said to be first or second associates, respectively).
(d) There exists a relationship of association between every pair of objects satisfying the following conditions:

(i) Each object has n_1 first and n_2 second associates.
(ii) Given any two objects that are ith associates, the number of objects common to the jth associates of the first and the kth associates of the second is p^i_{jk}, and this number is independent of the pair of objects with which we start. Furthermore, $p^i_{jk} = p^i_{kj}$ $(i, j, k = 1,2)$.

Clatworthy (1955 or 1973) has enumerated all such designs for which each object is replicated 10 or fewer times. These arrangements possess a high degree of symmetry and are therefore most suitable for incomplete paired-comparison experiments. Three main types of design may be distinguished, namely—

(1) *Group-divisible designs*

If t is expressible as the product of two integers m, n', both greater than 1, then the objects can be divided into m groups of size n'. Pairs are run λ_1 or λ_2 times, depending on whether or not they are in the same group. For example, in constructing a design for $t=6$, $r=3$ we may take $m=2$, $n'=3$, $\lambda_1=0$, $\lambda_2=1$ and write the objects as

$$\begin{matrix} 0 & 2 & 4 \\ 1 & 3 & 5 \end{matrix}$$

The comparisons made are easily written cyclically as $\{1, 3\}$. The argument generalizes directly to other basic $(\lambda_1+\lambda_2=1)$ GD designs and hence to all GD designs (Exercise 5.2). The GD designs with $\lambda_1=0$, $\lambda_2=1$ for $t \leqslant 14$ are listed in Table 5.1. Note that designs with $\lambda_1=1$, $\lambda_2=0$ are disconnected but may be useful in combination with the complete BIB design.

(2) *Triangular designs*

If $t=\frac{1}{2}n'(n'-1)$ an association scheme can be obtained by arranging the objects in order and symmetrically about the blanked-out principal diagonal of a $n' \times n'$ array. For example, if $t=6$, $n'=4$ the scheme is

$$\begin{matrix} * & 0 & 1 & 2 \\ 0 & * & 3 & 4 \\ 1 & 3 & * & 5 \\ 2 & 4 & 5 & * \end{matrix}$$

Two objects lying in the same row (or column) are first associates, whereas objects which do not lie in the same row are second associates. For $\lambda_1=1$, $\lambda_2=0$ we obtain the design:

01 02 12 03 04 34 13 15 35 24 25 45,

which is easily seen to be equivalent to the entry $t=6$, $r=4$ in Table 5.1 upon interchanging 3 and 5. Other basic triangular designs are given in Table 5.2 and these are all distinct from the group-divisible type.

Table 5.2 Triangular designs with $r \leqslant 10$, $\lambda_1 + \lambda_2 = 1$

From Clatworthy (1955) by permission of the author

Treatments t	Blocks b	Replications r	Size of array n'	λ_1	λ_2
6	12	4	4	1	0
10	15	3	5	0	1
10	30	6	5	1	0
15	45	6	6	0	1
15	60	8	6	1	0
21	105	10	7	1	0
21	105	10	7	0	1

(3) *Square designs*[*]

If t is the square of an integer $s>1$, an association scheme can be obtained by arranging the objects in order in a $s \times s$ square. Thus for $t=9$ we have

0 1 2

3 4 5

6 7 8

Here objects that are either in the same row or the same column are first associates, other objects are second associates. With $\lambda_1=1$, $\lambda_2=0$ we obtain the second design of Table 5.3, namely,

01 02 03 06 12 14 17 25 28

34 35 36 45 47 58 67 68 78

Further basic square designs are given in Table 5.3.

[*] Called Latin square designs in Clatworthy (1973).

Table 5.3 Square designs with $r \leqslant 10$, $\lambda_1 + \lambda_2 = 1$

From Clatworthy (1955) by permission of the author

Treatments t	Blocks b	Replications r	Size of square s	λ_1	λ_2
4	4	2	2	1	0
9	18	4	3	1	0
16	48	6	4	1	0
16	72	9	4	0	1
25	100	8	5	1	0
36	180	10	6	1	0

Triangular and square designs are not in general cyclic. Comparison with John *et al.* (1972) shows that they are only occasionally more efficient than the best comparable cyclic design. The group-divisible designs are always best. There is also a small class of cyclic PBIB(2) designs that are not group-divisible, with $(t, r) = (5, 2), (13, 6), (17, 8)$ (Clatworthy, 1955 or 1973). Only the first has maximal efficiency. Finally, Clatworthy (1973) lists seven PBIB(2) designs that do not fall into the main classes, with $(t, r) = (16, 5), (16, 6), (16, 9), (16, 10)$(twice), $(26, 10), (27, 10)$. These designs may or may not be maximally efficient. In none of the above cases is the efficiency of the best cyclic design more than 0·01 below the best PBIB(2) design.

John (1967) has made various modifications to GD designs to obtain three new classes of designs (some members of which are also cyclic). For example, his Type A designs are derived from GD's by systematically omitting certain comparisons. The resulting designs may not have a high degree of balance but in a few cases are more efficient(*) than cyclic or PBIB(2) designs of the same size.

5.5 Linked paired-comparison designs

If there are several judges and if one of the aims of the paired-comparison experiment is to measure agreement between them, then it is natural to require that the experiment should in some sense be balanced by judges as well as by comparisons. To this end

(*) In John (1967) and David (1963) efficiency figures need to be multiplied by $t/2(t-1)$ to convert them to values of E in (5.3.2).

each pair of judges must have certain comparisons in common. Consider, for example, a case of 4 objects, where the comparisons are difficult to make. If we assign 4 pairs to each judge, we need $n=3$ (or a multiple of 3) judges to obtain balance and might allocate the pairs as follows:

(5.5.1)	Judge	02	21	13	30
	Judge a	02	21	13	30
	b	01	12	23	30
	c	01	13	32	20

The rows are GD designs with $m=2$, $n'=2$, $\lambda_1=0$, $\lambda_2=1$ coming from the groupings

$$01, 23 \qquad 02, 13 \qquad 03, 12.$$

The result is an experiment in which every judge deals with a highly balanced set and every pair of judges have two different comparisons in common. However, requirements on the pairs compared by a single judge have to be relaxed in general to achieve overall balance.

Bose (1956) has considered the subject systematically, and the above is actually the first of his designs reproduced in Table 5.4. Let r' be the number of pairs of objects compared by each judge. To ensure symmetry between objects and judges, Bose introduces *linked paired-comparison designs* which require that:

(a) Among the r' pairs, compared by each of the n judges, each object appears equally often, say α times.

(b) Each pair is compared by k judges, $k>1$.

(c) Given any two judges there are exactly λ pairs which are compared by both judges.

Thus for experiment (5.5.1) we have

$$t=4, \quad n=3, \quad r'=4, \quad \alpha=2, \quad k=2, \quad \lambda=2, \quad b=6,$$

where $b=\frac{1}{2}t(t-1)$ is the number of possible pairs.

Quite generally it is clear that

$$r'=\tfrac{1}{2}t\alpha. \tag{5.5.2}$$

There is a certain rather subtle correspondence between linked paired-comparison designs and balanced incomplete block designs. Each judge may be considered to correspond to a treatment, and each pairing to a block. If a pair is compared by a judge, then the block corresponding to the pair may be considered to contain the treatment corresponding to the judge. Fig. 5.4

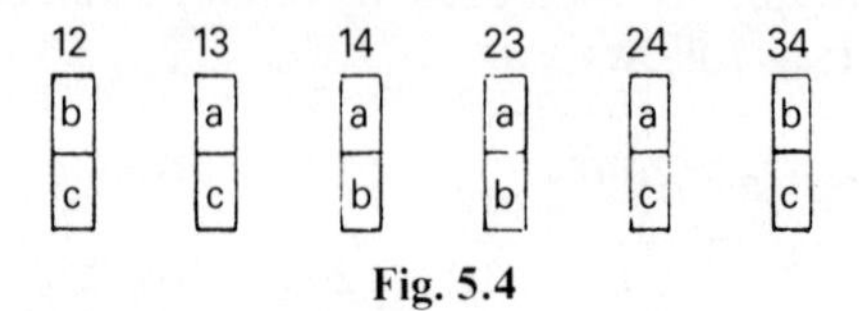

Fig. 5.4

illustrates the connexion for (5.5.1). Hence if a linked paired-comparison design of the type considered exists there must exist a corresponding BIB design with n treatments and b blocks, such that each block contains k treatments, each treatment occurs in r' blocks, and two given treatments occur together in λ blocks. It follows at once that

$$bk = nr', \quad \lambda(n-1) = r'(k-1). \tag{5.5.3}$$

Substituting for b and r' we can write the first of these equations as

$$k(t-1) = n\alpha.$$

It should be remembered that the existence of a BIB design with parameters n, $b = \frac{1}{2}t(t-1)$, r', k, λ does not ensure the existence of a corresponding linked paired-comparison design due to the additional restriction (a). For the construction of Designs 2, 3 of Table 5.4 (and a large design for $t=9$) we refer the reader to Bose's paper. We give now his very general method of deriving linked paired-comparison designs from BIB designs. First suppose the number of objects to be even, say $t=2i$. Then we can divide the $i(2i-1)$ pairs into $2i-1$ sets of i pairs each, such that every object occurs exactly once among the pairs of a set. For example, if $t=6$, we can take the objects to be 0, 1, 2, 3, 4, ∞. Then the 5 sets are

I	14	23	0∞
II	20	34	1∞
III	31	40	2∞
IV	42	01	3∞
V	53	12	4∞

Table 5.4 Linked paired-comparison designs

From Bose (1956) by permission of the author and the Editor of *Biometrika*

No.	Parameters	Design						
		Judge	Pairs assigned to a judge					
1	$t=4, n=3$	a	03	02	13	12		
	$b=6, r'=4$	b	02	13	01	23		
	$k=2, \lambda=2$	c	03	01	12	23		
	$\alpha=2$							
2	$t=5, n=6$	a	24	13	02	03	14	
	$b=10, r'=5$	b	12	23	03	04	14	
	$k=3, \lambda=2$	c	12	24	01	34	03	
	$\alpha=2$	d	24	01	23	13	04	
		e	01	23	34	02	14	
		f	12	34	13	02	04	
3	$t=6, n=10$	a	01	02	12	34	35	45
	$b=15, r'=6$	b	12	13	23	04	15	45
	$k=4, \lambda=2$	c	02	03	23	14	15	45
	$\alpha=2$	d	03	13	01	24	25	45
		e	03	05	35	12	14	24
		f	02	04	24	13	15	35
		g	01	04	14	23	25	35
		h	03	04	34	12	15	25
		j	02	05	25	13	14	34
		k	01	05	15	23	24	34

(continued overleaf)

Table 5.4 (*continued*)

No.	Parameters	Sets of pairs	Design	
			Judge	Sets of pairs assigned to a judge
4	$t=6, n=5$	I 03 12 45	a	II, III, IV, V
	$b=15, r'=12$	II 14 23 05	b	I, III, IV, V
	$k=4, \lambda=9$	III 20 34 15	c	I, II, IV, V
	$\alpha=4$	IV 31 40 25	d	I, II, III, V
		V 42 01 35	e	I, II, III, IV
5	$t=7, n=3$	I 01 12 23 34	a	II, III
	$b=21, r'=14$	45 56 60	b	III, I
	$k=2, \lambda=7$	II 02 13 24 35	c	I, II
	$\alpha=4$	46 50 61		
		III 03 14 25 36		
		40 51 62		
6	$t=8, n=7$	I 05 14 23 67	a	I, V, VII
	$b=28, r'=12$	II 16 25 34 07	b	II, VI, I
	$k=3, \lambda=4$	III 20 36 45 17	c	III, VII, II
	$\alpha=3$	IV 31 40 56 27	d	IV, I, III
		V 42 51 60 37	e	V, II, IV
		IV 53 62 01 47	f	VI, III, V
		VII 64 03 12 57	g	VII, IV, VI
7	$t=8, n=7$		a	III, V, VI, VII
	$b=28, r'=16$	as for No. 6	b	IV, VI, VII, I
	$k=4, \lambda=8$		c	V, VII, I, II
	$\alpha=4$		d	VI, I, II, III
			e	VII, II, III, IV
			f	I, III, IV, V
			g	II, IV, V, VI

In the general case, the $2i-1$ sets can be obtained by developing, mod $(2i-1)$, the initial set

$$1, 2i-2 \quad 2, 2i-3 \quad \ldots \quad i-1, i \quad 0\infty,$$

the object ∞ remaining unchanged.

Let us now take a BIB design with n_1 treatments, $b_1 = 2i-1$ blocks, r_1 replications, block size k_1, and in which every pair of treatments occurs together in the same block λ_1 times, and make each block correspond to one set and each treatment correspond to one judge. We can then obtain a linked paired-comparison design by assigning to each judge the sets of pairs corresponding to all blocks in which the treatment corresponding to the judge occurs. We obtain in this way a linked paired-comparison design with parameters

$$(5.5.4) \qquad t=2i, \quad n=n_1, \quad b=i(2i-1), \quad r'=ir_1, \quad k=k_1, \quad \lambda=i\lambda_1, \quad \alpha=r_1.$$

In the case $t=6$, we may start with the BIB design having parameters

$$n_1=5, \quad b_1=5, \quad r_1=4, \quad k_1=4, \quad \lambda_1=3.$$

If the treatments are taken as a, b, c, d, e, then the blocks are

a b c d

b c d e

c d e a

d e a b

e a b c

Since a occurs in lines 1, 3, 4, 5 we assign sets I, III, IV, V to judge a and obtain the following linked paired-comparison design

Judge	Sets			
a	I	III	IV	V
b	I	II	IV	V
c	I	II	III	V
d	I	II	III	IV
e	II	III	IV	V

The parameters of the design are by (5.5.4)

$$t=6, \quad n=5, \quad b=15, \quad r'=12, \quad k=4, \quad \lambda=9, \quad \alpha=4.$$

This is Design 4 of Table 5.4, except for some renumbering of the objects.

If the number of objects is odd, say $t=2i+1$, the $i(2i+1)$ pairs may be divided into i cyclic sets of $2i+1$ pairs each, as described in §5.3. We now take a BIB design with parameters n_2, $b_2=i$, r_2, k_2, and λ_2. Proceeding as before, we obtain a linked paired-comparison design with parameters

(5.5.5) $\quad t=2i+1, \quad n=n_2, \quad b=i(2i+1), \quad r'=(2i+1)r_2,$
$$k=k_2, \lambda=(2i+1)\lambda_2, \quad \alpha=2r_2.$$

In the case $t=7$ the BIB design consisting of the 3 comparisons of 3 objects gives Design 5 of Table 5.4.

To sum up, we see that every BIB design can be converted into a linked paired-comparison design: if the number of blocks in the former is even we can use (5.5.5); if the number is odd we have the choice of (5.5.4) and (5.5.5). From a list of BIB designs (e.g., Fisher

Table 5.5 Linked paired-comparison designs for $t \leqslant 20$ derivable by (5.5.4) or (5.5.5) from BIB plans in Cochran and Cox (1957).

t	n	b	r'	k	λ	α	Plan
4	3	6	4	2	2	2	3 × 2(*)
6	5	15	12	4	9	4	5 × 4(*)
8	7	28	12	3	4	3	11·7
8	7	28	16	4	8	4	11·8
8	7	28	24	6	20	6	7 × 6(*)
9	4	36	27	3	18	6	4 × 3(*)
10	9	45	40	8	35	8	9 × 8(*)
11	5	55	44	4	33	8	5 × 4(*)
12	11	66	30	5	12	5	11·19
12	11	66	36	6	18	6	11·20
12	11	66	60	10	54	10	11 × 10(*)
13	4	78	39	2	13	6	11·1
13	6	78	65	5	52	10	6 × 5(*)
14	13	91	28	4	7	4	11·22
14	13	91	63	9	42	9	11·23
15	7	105	45	3	15	6	11·7
15	7	105	60	4	30	8	11·8
15	7	105	90	6	75	12	7 × 6(*)
16	6	120	40	2	8	5	11·3
16	6	120	80	4	48	10	11·6
16	10	120	48	4	16	6	11·16
16	10	120	72	6	40	9	11·18
16	15	120	56	7	24	7	11·25
16	15	120	64	8	32	8	11·26
17	8	136	119	7	102	14	8 × 7(*)
19	9	171	152	8	133	16	9 × 8(*)
20	19	190	90	9	40	9	11·31
20	19	190	100	10	50	10	11·32

(*) The BIB design consists of all possible combinations of n in groups of k.

and Yates, 1957, Table XVIII, or Cochran and Cox, 1957, Table 9.5) many linked paired-comparison designs can be constructed. Designs 1 and 4–7 of Table 5.4 are a few of those obtainable in this way (Exercise 5.4). A list for $t \leqslant 20$ is given in Table 5.5.

Some of these designs were considered by Kendall (1955), who also discusses less well balanced arrangements.

EXERCISES

5.1 Show that for $t=17$, $r=4$ there are four nonisomorphic cyclic designs and find these.

5.2 Show that group-divisible 2-associate paired-comparison designs are cyclic.

5.3 Let $n_{ij\gamma}=1$, if A_i and A_j are compared by judge γ
$=0$, otherwise (including $n_{ii\gamma}=0$)
$(i, j=1, 2, \ldots, t, i \neq j; \gamma=1, 2, \ldots, n)$.
Then $N_\gamma=(n_{ij\gamma})$ may be called the *incidence matrix* for judge γ.
Show that necessary and sufficient conditions for the matrices N_γ to specify a linked paired-comparison design are

(a) $N_\gamma E=\alpha E \quad (\gamma=1, 2, \ldots, n)$,

(b) $\sum_{\gamma=0}^{n} N_\gamma=kE$,

(c) $\operatorname{trace}(N_\gamma N_{\gamma'})=2\lambda \quad (\gamma, \gamma'=1, 2, \ldots, n; \gamma \neq \gamma')$,

where E is a square matrix of order t, each of whose elements is unity, and $N_0=kI$, I being the identity matrix of order t.
(Wilkinson, 1957)

5.4 Show how Design 1 of Table 5.4 may be derived by the general approach of §5.5. Also obtain Designs 7 and 8 from Plans 11.7 and 11.8 in Cochran and Cox.

6 SELECTION AND RANKING

6.1 Ranking from Round Robin tournaments

We have repeatedly noted the analogy between a balanced paired-comparison experiment involving t objects and a Round Robin tournament of t players. In this chapter we use the language of the two situations interchangeably. Round Robins provide a natural means for producing a scoring, and hence a ranking, of the players; so far in this monograph three major approaches have been presented. In Chapter 3 several simple nonparametric multiple comparison procedures based on the row–sum scores are developed. Slater's nearest adjoining order of §2.2, resting on weak stochastic transitivity assumptions, is purely a ranking. Estimates of the merits in the linear models of Chapter 4 furnish parametric scores, and in the case of the Bradley–Terry model result in the same ranking as do the row–sum scores.

Each of the three approaches is usually preceded by a test of randomness. If this is not significant at the chosen level, one proceeds no further. Otherwise one uses a nearest adjoining order as a ranking with Slater's method but has some additional controls on significance levels with the first and third methods.

Sometimes a ranking is required regardless of questions of significance. This is particularly the case for top-scorers, where the problem of ranking merges with that of selecting the best object or possibly the best few objects. Before concentrating on the latter topic, we will examine a method that is intermediate, in the sense of providing a complete scoring of the objects, but placing emphasis on the top scores.

Let us reconsider the basic case of a simple strong tournament in which t players are ranked by their (row-sum) scores a_i. Since $a_i = t-1$ is impossible, as is $a_i = 0$, there must be at least two tied pairs or a tied triple of players. If a decision between the tied players must be made, and the time-honoured practice of a play-off is not feasible, it becomes important to develop tie-breaking methods.

The Kendall–Wei approach

One simple procedure, long used in chess tournaments, is to replace a_i by the sum of the scores of players defeated by A_i. To understand generalizations of this approach, mainly due to Kendall (1955) and Wei (1952), it is useful to regard the tournament outcome as a $t \times t$ matrix. An example is given in Table 6.1,

Table 6.1 Various score vectors for the tournament with matrix A

	A	**a**	$\mathbf{a}^{(2)}$	$\mathbf{a}^{(3)}$	$\mathbf{a}^{(4)}$	**s**
A_1	0 1 1 1 0	3	6	7	13	·6382
A_2	0 0 1 1 1	3	4	6	13	·5400
A_3	0 0 0 1 1	2	2	4	9	·3415
A_4	0 0 0 0 1	1	1	3	6	·2159
A_5	1 0 0 0 0	1	3	6	7	·3712

where, e.g., the score $a_1^{(2)}$ of A_1 at the second stage is $3+2+1=6$. Note that not only have all ties been broken at this stage but A_5 has moved ahead of A_3 although $a_5 < a_3$. Neither of these events (breaking of all ties, reversal of some rankings) need happen and one may wish to continue the re-allocation process to obtain $\mathbf{a}^{(3)}$, etc. We see that at the rth stage ($r=1, 2, \dots$) $\mathbf{a}^{(r)} = \mathbf{A}^r\mathbf{1}$, where $\mathbf{1}$ is the column vector of t 1's.

Now for $t > 3$ the matrix $\mathbf{A}$ of any strong tournament is *primitive*, i.e., $\mathbf{A}^r$ has all its elements positive from a certain integer $r = r_0$ on (Thompson, 1958; Moon, 1968, p. 35). It follows from Perron–Frobenius theory (e.g., Seneta, 1973) that

$$\lim_{r\to\infty} \left(\frac{\mathbf{A}}{\lambda}\right)^r \mathbf{1} = \mathbf{s},$$

where λ is the unique positive eigenvalue of $\mathbf{A}$ with the largest absolute value and $\mathbf{s}$ is a vector of positive terms. Here $\mathbf{s}$ is the column eigenvector satisfying

$$\mathbf{A}\mathbf{s} = \lambda\mathbf{s}, \tag{6.1.1}$$

and is determined only up to a constant multiplier.(*) In Table 6.1 $\mathbf{s}$ has been normalized ($\mathbf{s}'\mathbf{s} = 1$) and happens to give the same

(*) For a method of determining λ and $\mathbf{s}$ see, e.g., Searle (1982, p. 293).

ranking as $\mathbf{a}^{(2)}$. Note that these procedures are equally applicable to replicated Round Robins.

Interesting related methods, mostly based on the concept of "fair scores", have been put forward by Daniels (1969) and Moon and Pullman (1970). For example, suppose that A_i is "worth" V_i in the sense that any player who beats A_i wins V_i from A_i. One way of choosing the V_i is to equate expected gains and losses; i.e., we require

$$\sum_j \pi_{ij} V_j = V_i \sum_j \pi_{ji} \quad \mathrm{i} = 1, 2, \ldots, t,$$

where $\pi_{ii} = 0$ by our convention.

In a Round Robin of n rounds, π_{ij} is estimated by α_{ij}/n. Correspondingly we may estimate V_i by scores v_i satisfying

$$\sum_j \alpha_{ij} v_j = v_i \sum_j \alpha_{ji}, \tag{6.1.2}$$

or, defining

$$q_{ij} = \alpha_{ij} \Big/ \sum_j \alpha_{ji}, \quad \text{and} \quad \mathbf{Q} = (q_{ij}), \text{ by } \mathbf{Qv} = \mathbf{v}.$$

This is, in fact, the characteristic equation corresponding to $\mathbf{Q}$ which is a stochastic matrix with column sums 1, and hence has maximum eigenvalue 1. See also Exercise 6.3. For the eight non-isomorphic tournaments existing for $t \leqslant 5$ $(n = 1)$, $\mathbf{v}$ produces the same rankings as $\mathbf{s}$ of (6.1.1) except for breaking one additional tie (David, 1971).

It is dubious whether the rankings corresponding to $\mathbf{a}^{(2)}$, $\mathbf{s}$, or $\mathbf{v}$ are really an improvement over the row–sum score ranking $\mathbf{a}$. The common feature of the three methods is to give more credit to a player for defeating a high-scoring than a low-scoring opponent, but this means, of course, that a loss to the latter is punished less than a loss to the former. Also it is easy to show that an interchange of wins and losses does not necessarily reverse a ranking. Moreover, Rubinstein (1980) has proved that among procedures like the above (i.e., satisfying certain natural conditions) row–sum scoring is the only one for which the relative ranking of two players is independent of the outcome of matches in which neither is involved. Thus use of $\mathbf{a}^{(2)}$, $\mathbf{s}$, etc. is perhaps best reserved for breaking ties among high row–sum scores.

Kendall and Wei actually used $\mathbf{A}+\frac{1}{2}\mathbf{I}$ in place of the simpler $\mathbf{A}$ introduced by Moon (1968, p. 45). Variants of their method have been proposed by Thompson (1958), Hasse (1961), and Ramanujacharyulu (1964). See also Moon (1968, Topic 15) and David (1971). Finally, it is worth noting that draws, scored $\frac{1}{2}:\frac{1}{2}$, pose no extra complications in the ranking methods of this section, nor does a more general sharing of a point.

Unbalanced data

In dealing with balanced data Ramanujacharyulu (1964) considers powering $\mathbf{B}'$ rather than $\mathbf{B}$ to obtain the "iterated weakness" vector $(\mathbf{B}')^r\mathbf{1}$. The best player is now the one who has suffered the fewest iterated losses and is not necessarily the one who has scored the largest number of iterated wins. A balance between rewarding beating strong players and punishing losing to weak players is struck by the difference vector

$$\mathbf{d}^{(r)}=\mathbf{B}^r\mathbf{1}-(\mathbf{B}')^r\mathbf{1}. \qquad (6.1.3)$$

Consider now an incomplete tournament in which each pair of players has met at most once. It is seen immediately from (6.1.3) that

$$\sum_{i=1}^{t} d_i^{(r)}=\mathbf{1}'\mathbf{d}^{(r)}=0$$

and that scores are reversed when wins and losses are interchanged (i.e., $\mathbf{B}$ replaced by $\mathbf{B}'$). We recommend $\mathbf{d}^{(2)}$ for general use since it reduces to row–sum scoring in the case of a balanced tournament. In other words, $\mathbf{d}^{(2)}$, (unlike $\mathbf{d}^{(r)}$ for $r>2$) cannot be used as a tie-breaker for a balanced tournament, which makes it more attractive in the absence of balance.

To establish the equivalence of $\mathbf{d}^{(2)}$ and $\mathbf{a}$ for a balanced tournament, note that $\mathbf{d}^{(2)}$ may alternatively be written as

$$\mathbf{d}^{(2)}=\mathbf{a}^{(2)}-\mathbf{c}^{(2)}+\mathbf{a}-\mathbf{c}, \qquad (6.1.4)$$

where $\mathbf{c}=\mathbf{A}'\mathbf{1}$ and $\mathbf{c}^{(2)}=\mathbf{A}'\mathbf{c}$. We have

$$d_i^{(2)}=\Sigma_1 a_j-\Sigma_2(t-1-a_j)+a_i-(t-1-a_i) \quad i=1,2,\ldots,t,$$

where $\Sigma_1(\Sigma_2)$ denotes summation over players A_j for whom $A_i \to A_j$ $(A_j \to A_i)$. (A tie is half a win plus half a loss.) But

$$\Sigma_1 a_j + \Sigma_2 a_j = \tfrac{1}{2}t(t-1) - a_i,$$

so that

$$\begin{aligned} d_i^{(2)} &= \tfrac{1}{2}t(t-1) - a_i - (t-1)(t-1-a_i) + 2a_i - (t-1) \\ &= t[a_i - \tfrac{1}{2}(t-1)]. \end{aligned}$$

In words, $d_i^{(2)}$ is the total number of (a) wins of players defeated by A_i minus losses of players to whom A_i lost, plus (b) A_i's wins minus A_i's losses. Clearly, (b) could be omitted without changing any of the preceding properties. However, in its absence, A_i after beating an opponent with no wins would be worse off than before, since the win adds nothing to part (a) of A_i's score but adds 1 to the score of each player who defeated A_i.

For a worked example see David (1987).

Formula (6.1.4) may also be used in larger tournaments. If A_i beats A_j α_{ij} times in $n_{ij}(\geqslant 1)$ encounters, then the (i, j) element of $\mathbf{A}$ becomes $p_{ij} = \alpha_{ij}/n_{ij}$; if $n_{ij} = 0$ both p_{ij} and p_{ji} are zero. It will be realized, however, that the present method, or any other non-parametric method, cannot be entirely satisfactory when the n_{ij} differ greatly. An alternative but less simple approach is given by least squares, as shown in §4.1. See also Cowden (1975) and Daley (1979).

6.2 Appropriate size of experiment

It will frequently be assumed in this and the following section that a linear model (§1.3) is applicable. A theoretical ranking of the objects may then be made according to their "merits". We associate object $A_{(i)}$ with merit $V_{(i)}$, where

$$V_{(1)} \leqslant V_{(2)} \leqslant \ldots \leqslant V_{(t)},$$

and write $a_{[i]}$ for the score of $A_{(i)}$. It is to be noted that $a_{[i]}$ is a theoretical quantity (not to be confused with $a_{(i)}$), as in a practical situation we do not know which object to label $A_{(i)}$. For convenience of writing we will, in this and the next section, leave off brackets on the subscripts of π's and will mean by π_{ij} the preference probability $\Pr\{A_{(i)} \to A_{(j)}\}$.

Formulation of the problem

In many experiments designed to compare t objects (or treatments) the primary interest lies in the detection of the best object. For a balanced paired-comparison experiment it is natural to declare as best the object with the highest score, but this may yield an incorrect result, due to chance fluctuations. However, if $A_{(t)}$ is strictly better than $A_{(t-1)}$, and if the number n of replications is large enough, then $A_{(t)}$ should emerge with the highest score with a probability P as close to 1 as desired. We shall now consider how n might be determined.

A possible procedure is as follows:

(a) Find n corresponding to given values of t, P, and the configuration of preference probabilities $C(\pi_{ij})$.

(b) Perform the experiment and declare best the object with the highest score; if m scores tie for first place declare best one of the corresponding objects at random.

Step (b) is straightforward, the selection rule being quite naïve so as to simplify the later discussion. However, (a) needs some elaboration. Since, for fixed t and P, n depends on $C(\pi_{ij})$ we shall specialize the π_{ij} in line with the general approach due to Bechhofer (1954). We suppose that $A_{(t)}$ has probability $\pi(>\frac{1}{2})$ of being preferred to each of the other objects and that the remaining $t-1$ objects are of equal value. This assumption, a special case of a linear model, expresses in simplified form that a superior object is present and would appear to be reasonable as a basis of determining the number of replications of the experiment. While completely analogous to the kind of specialization employed by Bechhofer, it turns out that in the present case this choice of the π_{ij} does not necessarily correspond to a least favourable case. n is given in Appendix Table 4, further discussed below.

The model chosen can be summed up thus:

$$\pi_{tj} = \pi > \tfrac{1}{2}, \quad j = 1, 2, \ldots, t-1; \tag{6.2.1}$$

$$\pi_{ij} = \tfrac{1}{2}, \quad i, j = 1, 2, \ldots, t-1; \quad i \neq j.$$

Exact distribution theory

From (2.4.3) and (2.4.4) the joint distribution of scores under model (6.2.1) may be written

(6.2.2) $$f(\mathbf{a}_{[i]}) = 2^{-n\binom{t-1}{2}}\ g(\mathbf{a};n)\,\pi^{a_{[t]}}(1-\pi)^{n(t-1)-a_{[t]}}.$$

Suppose that m scores tie for first place. Of the $G(\mathbf{a};n)/g(\mathbf{a};n)$ permutations of the scores $a_1, a_2, \ldots, a_t$, a proportion m/t must have a top score in the last position, that is, associated with $A_{(t)}$. With the randomization of step (b), the corresponding contribution to the probability of correct selection P is

(6.2.3) $$\frac{1}{m}.\ 2^{-n\binom{t-1}{2}}\ \frac{m}{t}\ G(\mathbf{a};n)\,\pi^{a_{[t]}}(1-\pi)^{n(t-1)-a_{[t]}},$$

which is independent of m. P is then given by summing (6.2.3) over all $a_{[t]}$ which can be maximum scores and over all permissible values of the other scores, and may be expressed as

(6.2.4) $$P = 2^{-n\binom{t-1}{2}} \sum_{a_{[t]}=c}^{n(t-1)} \pi^{a_{[t]}}(1-\pi)^{n(t-1)-a_{[t]}} \times \sum_{\sum_{i=1}^{t-1} a_{[i]} = n\binom{t}{2} - a_{[t]}} \frac{1}{t}\, G(a_1, a_2, \ldots, a_t; n),$$

where c is the smallest integer greater than or equal to $\frac{1}{2}n(t-1)$. E.g., if $t=3$ and $n=1$, (6.2.4) reduces to

$$P = 2^{-1}[\pi(1-\pi)\tfrac{1}{3}\ G(1,1,1;1) + \pi^2 \tfrac{1}{3}\ G(0,1,2;1)]$$

where $G(0, 1, 2; 1)$, the frequency of the partition [210], is 6, and $G(1, 1, 1; 1) = 2$. Thus, as is otherwise obvious in this simple case,

$$P = \pi^2 + \tfrac{1}{3}\ \pi(1-\pi).$$

The leading term of P is always $\pi^{n(t-1)}$.

The case $t=2$ has received special attention by Mosteller (1952) in the context of the "World Series" where two finalist U.S. baseball teams play until a series of 7 games is decided. Clearly, the fact that the series may terminate before all games have been played does not alter P, the probability that the better team will win. On the other hand, it is no longer satisfactory to estimate the probability that a chosen team will win an individual game, by the proportion $a_1/(a_1+a_2)$ of victories in its favour. We have here an extension of inverse binomial sampling since the series concludes when either a_1 or a_2 reaches $c=4$. The appropriate unbiased estimate is (Girshick, Mosteller, and Savage, 1946) $(c-1)/(c+a_2-1)$ or $a_1/(a_1+c-1)$, according as the chosen or the other team wins. Various practical complications, not entirely incomprehensible to non-Americans, are discussed by Mosteller. See also Exercise 6.1.

Comments on Appendix Table 4. For an experiment involving t objects with preference probabilities satisfying (6.2.1) the table gives the smallest number of replications n which ensures that the highest score in an experiment of size (t, n) will correspond to the best object with at least a preassigned probability P'. In the con-

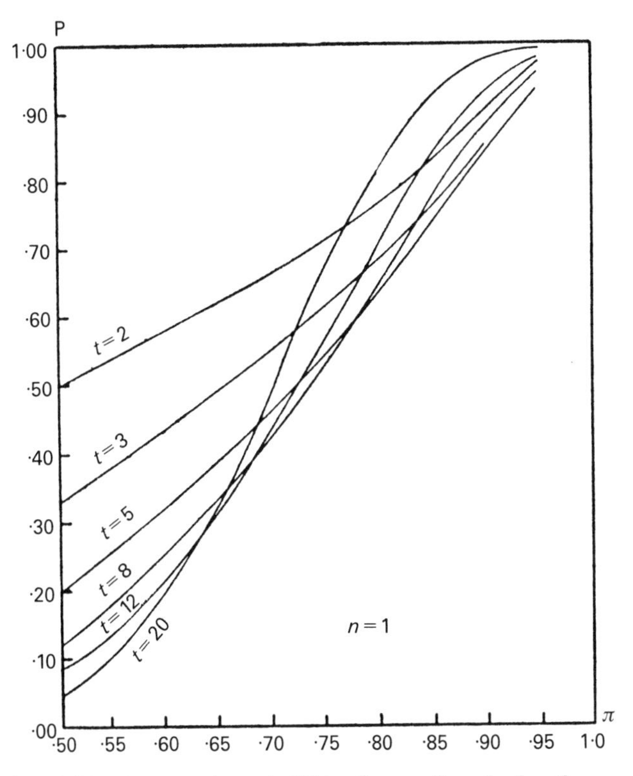

Fig. 6.1A, 6.1B Asymptotic probability of correctly selecting the superior object $A_{(t)}$ in the presence of $t-1$ equal others; n = number of replications, $\pi = \Pr(A_{(t)} \rightarrow A_{(i)})$, $i = 1, 2, \ldots, t-1$

struction of the table exact theory was used for the combinations (t, n) up to (2, 268), (3, 21), (4, 8), (5, 4), (6, 1), (7, 1), and (8, 1). Elsewhere asymptotic approximations derivable from the asymptotic theory of §2.4 are listed (see Trawinski and David, 1963). Comparisons of exact and asymptotic values showed good agreement for the larger of the above experiment sizes.

If model (6.2.1) were least favourable, the value of n obtained from the table would ensure correct selection with probability

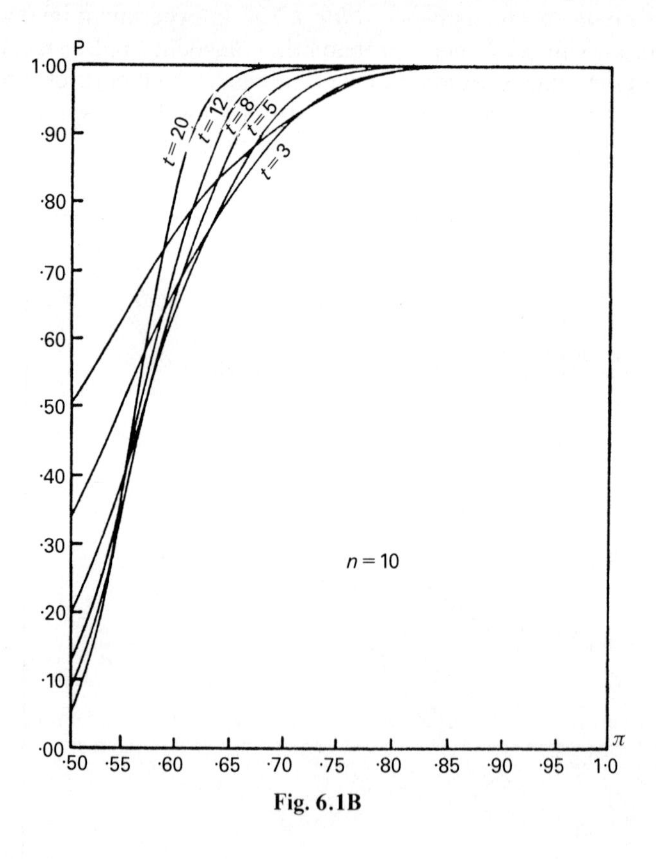

Fig. 6.1B

$P \geqslant P'$ as long as

(6.2.5) $$\Pr\{A_{(t)} \rightarrow A_{(t-1)}\} \equiv \pi_{t,t-1} \geqslant \pi.$$

That this is not necessarily so is most readily seen from Fig. 6.1A and 6.1B which show how P increases as a function of π for $n=1$

and $n=10$ when model (6.2.1) holds. For example, take $t=20$ and suppose that the first five objects, considered by themselves, satisfy the model (6.2.1) with $t=5$. Complete the specification by adding to these five objects fifteen of no value, that is, having probability zero of being preferred to any of the first five. Condition (6.2.5) is clearly satisfied in this augmented case. Now for $n=1$, Fig. 6.1A shows that P is greater for $t=20$ than for $t=5$ if $0{\cdot}69<\pi<0{\cdot}95$. Hence for this range of π the augmented case, which leads to the same value of P as the case $t=5$ graphed, is actually less favourable than the case labelled $t=20$. There is, of course, no guarantee that it is least favourable. In fact, considerations of this kind indicate that it would in general be difficult to determine the least favourable configuration and that this configuration may be far removed from a realistic situation. A conjecture along these lines is made by Gibbons, Olkin, and Sobel (1977, p. 226). Thus Appendix Table 4 can be used as a safe guide to the appropriate n only if model (6.2.1) holds. However, this model is of considerable importance in its own right, corresponding as it does to the situation of a single "outlier".

It may be of interest to re-phrase the above example in the language of tournaments. If a player A_i has probability $\pi(0{\cdot}69<\pi<0{\cdot}95)$ of defeating any one of 19 opponents, who are all of equal strength, then A_i has a better chance of winning a simple Round Robin tournament against all 19 opponents than of winning a smaller tournament against only four of these opponents. However, Huber (1963a) has shown that, for any $\pi>\frac{1}{2}$, the probability of the superior player's winning the tournament tends to 1 as $t\to\infty$.

Curtailed Round Robin

If the aim of the Round Robin is purely to pick the best player, curtailment of the tournament becomes possible. Clearly, the probability of a correct decision is unchanged if the tournament is stopped as soon as one player has reached a score that cannot be equalled by any other player. Such *weak curtailment* can result in an appreciable reduction in matches played. A further reduction is brought about by *strong curtailment* under which the tournament is stopped as soon as one player has attained a score that cannot be overtaken by any other player. If we imagine the tournament run to completion, the ultimate winner may be a player who drew equal to the early winner, since ties are broken at random. However, it can be shown that the two probabilities of correct selection are equal if the Bradley–Terry model holds, which includes the one-outlier model as as very special case.

The idea of strong curtailment was introduced by Bechhofer and Kulkarni (1982) in the context of binomial populations. In analogy to their modified experimental procedure for selecting the best binomial population, one can modify the running of the tournament by allowing a player, after a win, to go on playing a fresh opponent. This is, of course, possible only if matches can be scheduled one at a time, a situation more natural to paired-comparison experiments than tournaments. It is also possible to extend the procedure to the selection, with or without ranking, of more than one player. For details see David and Andrews (1987).

6.3 Selection of a subset containing the best object

Consider the set of objects $A=\{A_1, A_2, \ldots, A_t\}$ and let S be a subset of A consisting of those objects with the highest scores. In this section our aim is to select S just large enough to ensure, with at least a pre-assigned probability P^*, that the best object $A_{(t)}$ is included in S. Following Gupta and Sobel (1960) we use the decision rule $\mathscr{R}$:

Retain in S only those objects A_i for which $a_i \geqslant a_{(t)} - \nu$, where $a_{(t)}$ is the highest score and ν, a non-negative integer, is a function of t, n, and P^*.

For $P^* = 0{\cdot}75$, $0{\cdot}90$, $0{\cdot}95$, $0{\cdot}99$ and a wide range of values of t and n, the value of ν, determined from both exact and asymptotic theory, is given in Appendix Table 5. it should be noted that the size of S is a random variable which can range from 1 to t. The rule gives a correct selection if

$$a_{[t]} \geqslant a_{(t)} - \nu.$$

For fixed t and n, the probability of correct selection P_{CS} depends on ν and on the configuration of preference probabilities $C(\pi_{ij})$, and we have

$$P_{CS} = \Pr\{a_{[t]} \geqslant a_{(t)} - \nu \,|\, t, n, C(\pi_{ij})\}.$$

In the table ν has been chosen as the smallest integer making

$P_{CS} \geqslant P^*$ when $C(\pi_{ij}) = C(\frac{1}{2})$, i.e. when all objects are equivalent (we tag object $A_{(t)}$ so that $a_{[t]}$ is still defined). It can be shown (Trawinski and David, 1963) that for fixed t, n, ν, the configuration $C(\frac{1}{2})$ leads to the infimum of P_{CS} for any $C(\pi_{ij})$ satisfying a linear model. In this case we may speak of $C(\frac{1}{2})$ as a conservative configuration and have

$$P_{CS}\{C(\pi_{ij})\} \geqslant P_{CS}\{C(\tfrac{1}{2})\} \geqslant P^*.$$

Exact evaluation of ν

For the range of values of t and n covered by the tables of partitions referred to in §2.3 it is possible to determine ν exactly. We illustrate the procedure for $t=4$, $n=2$. In this case there are 16 permissible partitions of which a typical one is [5430] with frequency $G(5, 4, 3, 2; 0) = 114$. Since $t=4$ the frequency with which $a_{[t]}$ equals or exceeds $a_{(t)} - \nu = 5 - \nu$ must be a multiple m of $144/4 = 36$. We have for

$\nu = 0$, that $m = 1$, corresponding to $a_{[t]} = 5$,
$\nu = 1$, that $m = 2$, corresponding to $a_{[t]} = 5, 4$
$\nu = 2, 3, 4$ that $m = 3$, corresponding to $a_{[t]} = 5, 4, 3$
$\nu = 5$, that $m = 4$, corresponding to $a_{[t]} = 5, 4, 3, 0$.

Thus the contribution to $P_{CS}\{C(\frac{1}{2})\}$ from this partition is, for a given ν, $36m(\nu)/2^{12}$, the divisor being the factor converting frequencies into probabilities. If the corresponding contributions for all 16 partitions are added up, the resulting values of $P_{CS}\{C(\frac{1}{2})\}$ are:

ν	0	1	2	3	4	5
$P_{CS}\{C(\frac{1}{2})\}$	0·342 773	0·559 570	0·769 042	0·905 273	0·975 098	0·997 070

The entries in Appendix Table 5 for $t=4$, $n=2$ follow at once.

An example

We illustrate the application of the decision rule $\mathscr{R}$ on data supplied by E. Jensen of Faellesforeningen for Danmarks Brugsforeninger, Copenhagen. 15 persons examined all possible pairings of 4 different samples for taste. The following preference table was obtained:

	A_1	A_2	A_3	A_4	a_i
A_1	–	3	2	2	7
A_2	12	–	11	3	26
A_3	13	4	–	5	22
A_4	13	12	10	–	35

We have $a_{(t)}=35$. To ensure that with at least a preassigned probability $P^*=0{\cdot}75$ the best sample is in the selected subset we enter Appendix Table 5 for $t=4$, $n=15$, $P^*=0{\cdot}75$, find $\nu=7$ and hence retain only A_4 in the subset. For $P^*=0{\cdot}90$ we have $\nu=9$, so that the subset consists of A_4 and A_2, and so on.

Controlling the expected size of the subset

If the number of replications n is at our disposal, it is possible, at least in principle, to control the expected size of S for given $C(\pi_{ij})$. A natural basis for such design considerations is the one-outlier model (6.2.1), in which case the expectation may be denoted by $\mathscr{E}_{t,n}(\pi, \nu)$. Using both exact and asymptotic theory Trawinski (1985) has tabulated $t^{-1}\mathscr{E}_{t,n}(\pi, \nu)$ for $P^*=0{\cdot}75$, $0{\cdot}90$, $0{\cdot}95$, $0{\cdot}975$, $0{\cdot}99$, $t=3(1)20$, $n=1(1)20(5)50(10)100$, $\pi=0{\cdot}6(0{\cdot}1)0{\cdot}9$. Entering the table with arguments t and π, one can find the smallest value of n for which $\mathscr{E}_{t,n}(\pi, \nu)$ is below the specified upper bound, E^*, say. The table also provides the corresponding value of ν taken essentially from Appendix Table 5. Thus having found n and ν, we can say that under the one-outlier model $a_{[t]} \geqslant a_{(t)} - \nu$ with probability $\geqslant P^*$ *and* that $\mathscr{E}_{t,n}(\pi, \nu) \leqslant E^*$. Reference should be made to Trawinski (1985) for several examples illustrating the use of his table.

6.4 Knock-out and other tournaments

The analogy between balanced paired-comparison experiments and Round Robin tournaments suggests that other types of tournaments might also deserve closer scrutiny. A feature of such tournaments is that balance is given up either for speed, or for an increased number of encounters between the more successful players, or for both. Lack of balance increases the difficulty of studying the properties of these tournaments, but they have considerable intuitive appeal as a method of experimentation when our main purpose is the selection of the best object. Best known is the Knock-out (or Cup-tie) tournament which we discuss first.

Knock-out tournament

We suppose that the number t of players has been reduced, by preliminary matches or otherwise, to a power of 2, say $t=2^p$. A winner can then be declared after p rounds and $t-1$ games as against $t-1$ rounds and $\frac{1}{2}t(t-1)$ games required for a Round Robin tournament if no play-offs are necessary.

If a particular player A_1 beats any opponent with probability π, then A_1 will win the Knock-out tournament with probability π^p. In more complicated cases, A_1's winning probability will evidently depend on the initial tournament draw, which, to start with, we take to be strictly random. Let $P_i^{(r)}$ $(i=1, 2,\ldots,t;\ r=1, 2,\ldots,p)$ denote the probability that A_i reaches the rth round of the tournament *and* is victorious in this round, so that $P_i^{(p)}$ is A_i's probability of winning the tournament. Let $M_{ij}^{(r)}$ be the probability that A_i meets A_j $(j=1, 2,\ldots,t; j\neq i)$ in the rth round and M_{ij} be the probability of their meeting in the course of the tournament. Clearly

$$(6.4.1)\qquad M_{ij}=\sum_{r=1}^{p} M_{ij}^{(r)}.$$

Suppose now that

$$\pi_{1j}=\pi_1 \quad \pi_{2j}=\pi_2\ (j=3,\ldots,t).$$

Then A_1's probability of winning the tournament is

$$(6.4.2)\qquad P_1^{(p)}=M_{12}\pi_{12}\pi_1^{p-1}+(1-M_{12})\pi_1^p.$$

If the draw is known, M_{12} and hence $P_1^{(p)}$ can be evaluated. Instead of doing this, we shall obtain expressions for $\bar{M}_{12}$ and $\bar{P}_1^{(p)}$, the values of M_{12} and $P_1^{(p)}$ *before* the draw. These prior probabilities will still be linked by (6.4.2). We have

$$\bar{M}_{12}^{(1)} = 1/(t-1).$$

If A_1 and A_2 are to meet in the second round, in which there will be $\frac{1}{2}t$ contestants, they must both be successful in the first round, so that

$$\bar{M}_{12}^{(2)} = (1 - \bar{M}_{12}^{(1)})\pi_1\pi_2/(\tfrac{1}{2}t - 1)$$
$$= 2\pi_1\pi_2/(t-1).$$

Continuing the argument, we find

$$\bar{M}_{12}^{(r)} = (2\pi_1\pi_2)^{r-1}/(t-1),$$

and hence from (6.4.1)

$$\bar{M}_{12} = \sum_{r=1}^{p} (2\pi_1\pi_2)^{r-1}/(t-1).$$

It is interesting to note that the $\bar{M}_{ij}^{(r)}$ depend on π_1 and π_2 only through the product $\pi_1\pi_2$. We see also that the probability of a final between A_1 and A_2 is $\bar{M}_{12}^{(p)}$. If for the moment we imagine A_1 and A_2 to be "seeded" players placed in opposite halves of the draw, this probability would become

$$(\pi_1\pi_2)^{p-1} \sim 2\bar{M}_{12}^{(p)}.$$

To illustrate the general case with preference probabilities $C(\pi_{ij})$ take $t=8$ and suppose (without loss of generality) that $A_1, A_2, \ldots, A_8$ have been drawn in that order. Then we have, for example,

$$P_1^{(1)} = \pi_{12}, \quad P_2^{(1)} = \pi_{21}, \quad P_3^{(1)} = \pi_{34}, \ldots,$$

(6.4.3) $$P_1^{(2)} = P_1^{(1)}(\pi_{13}P_3^{(1)} + \pi_{14}P_4^{(1)}) = \pi_{12}(\pi_{13}\pi_{34} + \pi_{14}\pi_{43}),$$

(6.4.4) $$P_1^{(3)} = P_1^{(2)}(\pi_{15}P_5^{(2)} + \pi_{16}P_6^{(2)} + \pi_{17}P_7^{(2)} + \pi_{18}P_8^{(2)}).$$

The corresponding probabilities before the draw can be obtained by averaging over all possible draws. Let $P_1^{(2)}(i, j, k)$ denote the joint probability that A_1 beats A_i in the first round and the winner of A_j v. A_k in the second round; then $P_1^{(2)}(2, 3, 4)$ is the expression of (6.4.3). Likewise, we may (for $t=8$) write $P_1^{(3)}$ of (6.4.4) more fully as $P_1^{(3)}(2, 3, 4)$. It follows that

(6.4.5) $$\bar{P}_1^{(1)} = \tfrac{1}{7} \sum_{i=2}^{8} \pi_{ij},$$

$$\bar{P}_1^{(2)} = \tfrac{1}{105} \sum_{\substack{i \\ 1 \neq i \neq j \neq k}} \sum_{j<k} \sum P_1^{(2)}(i, j, k), \tag{6.4.6}$$

$$\bar{P}_1^{(3)} = \tfrac{1}{105} \sum_{\substack{i \\ 1 \neq i \neq j \neq k}} \sum_{j<k} \sum P_1^{(3)}(i, j, k).$$

For $t > 8$ the number of terms on the right-hand side may become formidable. Thus $\bar{P}_1^{(r)}$ is the mean of

$$(t-1)\binom{t-2}{2}\binom{t-4}{4}\cdots\binom{t-2r-1}{2^{r-1}} \text{ terms.}$$

Estimation of the strengths of players in a deterministic KO tournament

From the results of a KO tournament of size $t = 2^p$ it is possible to obtain estimates of the strengths of the players (Hartigan, 1966; Moon, 1970; Narayana and Agyepong, 1980). For convenience, label the players 1, 2,..., t in increasing order of strength. It is supposed here that a stronger player always beats a weaker one, the only random element in the tournament being the draw. Then the probability that player i survives round k is

$$s(i, k) = \binom{i-1}{2^k-1} \bigg/ \binom{2^p-1}{2^k-1}, \tag{6.4.7}$$

since the $2^k - 1$ other players that are in the ith part of the tournament during the first k rounds must all be inferior to i, i.e., drawn from the $i-1$ weakest players. Then

$$r(i, k) = s(i, k) - s(i, k+1)$$

is the probability that $i(<2^p)$ plays exactly $k+1$ rounds ($k = 0, 1, \ldots, p-1$). If R is the number of rounds played by i, then

$$\mathscr{E}(R) = \sum_{k=0}^{p-1} (k+1) r(i, k) = \sum_{k=0}^{p-1} s(i, k)$$

and

$$\mathscr{E}[R(R+1)] = 2 \sum_{k=1}^{p-1} ks(i, k).$$

Turning to the estimation of the strength X of a player beaten by i in round k, we note that this must be an integer, ν, satisfying $2^{k-1} \leqslant \nu < i$. Also, by (6.4.7), with $K = 2^{k-1}$,

$$p^*(\nu, k) = \frac{s(\nu, k-1)}{\sum_{\nu=K}^{i-1} s(\nu, k-1)} \tag{6.4.8}$$

is the probability that i beats ν in round k, given i has reached round k. Then

$$\begin{aligned} \mathscr{E}(X) &= \sum_{\nu=K}^{i-1} \nu p^*(\nu, k) \\ &= \sum_{\nu=K}^{i-1} \nu \binom{\nu-1}{K-1} \Big/ \sum_{\nu=K}^{i-1} \binom{\nu-1}{K-1} \\ &= K \binom{i}{K+1} \Big/ \binom{i-1}{K} = \frac{iK}{K+1}. \end{aligned}$$

Thus $\mathscr{E}(X)/i$ depends only on k and equals $\frac{1}{2}$ for $k=1$. Also $\operatorname{var}(X)$ is easily found since

$$\mathscr{E}[X(X+1)] = \frac{i(i+1)K}{K+1}.$$

Suppose now that i loses in turn to t in round l. Then by a continuation of the above argument, the expected strength of the player defeated by i in round k is

$$\frac{2^{k-1}}{2^{k-1}+1} \cdot \frac{2^{l-1}}{2^{l-1}+1} \cdot 2^p.$$

For a brief treatment of deterministic double and r-elimination tournaments ($r \geqslant 2$) see Fox (1973).

Random KO tournaments

Suppose the number of players is $t = 2^p + q (0 \leqslant q < 2^p)$. A random KO tournament is determined by the vector of positive integers $\mathbf{t} = (t_1, \ldots, t_k)$ satisfying

$$t_1 + \ldots + t_k = t - 1,\ t_k = 1$$
$$2t_1 \leqslant t$$
$$2t_i \leqslant t - t_1 - \ldots - t_{i-1} \quad (i \geqslant 2).$$

In the first round of the tournament $2t_1$ players are chosen at random and paired off randomly, the remaining $t - 2t_1$ players having byes. The t_1 losers are eliminated, leaving $t - t_1$ players, with vector $(t_2, \ldots, t_k)$. The process is then repeated, all $t - t_1$ players being treated equally, etc.

Random KO tournaments were introduced by Narayana (1968). This and subsequent work are summarized in Narayana (1979, chapter IV). Examples of random KO tournaments are

$$T_1: \mathbf{t}_1 = (u_1, u_2, \ldots, u_k), \quad \text{where}$$

$$u_i = \left[\frac{t + 2^{i-1} - 1}{2^i}\right],\ i = 1, 2, \ldots, k, \text{ and}$$

$[x]$ is the integral part of x.

$$T_2: \mathbf{t}_2 = (q, 2^{p-1}, 2^{p-2}, \ldots, 1),$$
$$T_3: \mathbf{t}_3 = (1, \ldots, 1).$$

In T_1 as many players as possible are paired in each round, whereas T_2 is the familiar (Wimbledon) method of pairing off $2q$ players in a preliminary round so as to reduce the number of players to a power of 2. For $q = 0$ both T_1 and T_2 reduce to the classical KO tournament. T_3 requires the maximum number of rounds, $t - 1$. An important modification is T_4 which has the same vector as T_3 but with the requirement that the winner of each round meets a random opponent in the next round. Such a "play the winner" tournament is not strictly of the random KO type.

For the single outlier model (6.2.1) Narayana (1979) finds for $\frac{1}{2} < \pi < 1$ and $t \geqslant 4$

$$P_2(\pi) \geqslant P_1(\pi) > P_3(\pi) > P_4(\pi),$$

where, e.g., $P_2(\pi)$ is the probability that the best player wins T_2.

Maurer (1975) proves that the winning probability of the best player is at least as large for T_2 as for any other tournament of $t - 1$ (or fewer) games. Moreover, he establishes this not just for the single-outlier model but more generally in the following two cases:

(a) $\pi_{tj} = \pi \geqslant \frac{1}{2}$ for $j = 1, \ldots, t-1$

(b) $\pi_{tj} \geqslant \frac{1}{2}, \pi_{ij} = \frac{1}{2}$ for $i, j = 1, \ldots, t-1, i \neq j.$

In (a) the π_{ij} for $i, j < t$ are left unspecified. This clearly does not affect $P_2(\pi)$ as obtained under (6.2.1) but the "best" player no longer necessarily has the highest probability of winning T_2.

Comparison of the effectiveness of tournaments

A simple measure of the effectiveness of a tournament is the probability that the best player (i.e., the one with the highest expected score) wins the tournament. For various configurations $C(\pi_{ij})$ of preference probabilities the Round Robin was compared with Knock-out type tournaments for $t = 4, 8$ by Glenn (1960) and Searls (1963). In order that the tournaments under comparison involved roughly equal numbers of games, ordinary KO tournaments were replicated in various ways. Later replications may be run so as to take advantage of earlier results.

This approach was extended to random KO and related tournaments by Narayana and co-workers (see Narayana, 1979) but with $C(\pi_{ij})$ specialized to the one-outlier model (6.2.1). It is obvious that the Round Robin will not fare well in such comparisons if the measure of effectiveness is the probability of correctly selecting just the best player; its relative effectiveness is bound to grow as we increase the number of players to be selected or ranked.

For $t = 3$ a "drop-the-loser" rule is appealing: After each game the loser is given a bye. A tournament winner may be declared as soon as (for pre-assigned integers c_1, c_2, c_3) (a) one player has a total of c_1 wins; (b) two players have suffered c_2 losses; (c) one player has established a lead of c_2. These alternatives are investigated under general $C(\pi_{ij})$ by Sobel and Weiss (1970) and Bechhofer (1970).

An interesting comparison of the effectiveness of tournaments in picking the best player has been made by Handa and Maitri (1984). They consider in detail two generalizations of T_2 above, namely $T_2(c)$ and $T_2(l, c)$. In both these tournaments Knock-out procedures are used but with a sequence of up to $2c-1$ matches replacing a single encounter. The winner of a sequence is declared in $T_2(c)$ as soon as one player has won c matches and in $T_2(l, c)$ as soon as one player leads the other by l wins *or* has won c matches. Note that $T_2(c) = T_2(c, c)$. That $T_2(c)$ provided a good performance in Glenn (1960), Narayana and Hill (1974), and Narayana (1979) motivated its refinement to $T_2(l, c)$.

Let $\rho_{ij}(l, c)$ be the probability that A_i knocks out A_j if they meet in $T_2(l, c)$. If a match won by A_i is represented by a unit step on the x-axis and a win by A_j by a unit step on the y-axis, then the results of the sequence of matches between A_i and A_j can be put in 1–1 correspondence with the set of minimal lattice paths starting from the origin and first reaching one of the boundary lines $y = x \pm l$, $x = c$, $y = c$. Because of this correspondence, it is seen that

$$\rho_{ij}(l, c) = \sum_{\nu=0}^{c-1} G_l(m-1, \nu)\pi_{ij}^{m}\pi_{ji}^{\nu}, \tag{6.4.9}$$

where $m = \min(l + \nu, c)$ and

$$G_l(m-1, \nu) = \text{number of paths from } (0, 0) \text{ to } (m-1, \nu) \text{ not touching the boundaries } y = x \pm l.$$

Also it is known (e.g., Mohanty, 1979, p. 6) that

$$G_l(m, \nu) = \sum_k \left[\binom{m+\nu}{m-2kl} - \binom{m+\nu}{\nu+2kl+l}\right],$$

where $\binom{a}{b} = 0$ if $b < 0$ or $a < b$. In particular, if $l = c$, then (6.4.9) reduces to

$$\rho_{ij}(c, c) = \pi_{ij}^{c} \sum_{v=c}^{2c-1} \binom{v-1}{c-1} \pi_{ji}^{v-c},$$

as is otherwise obvious.

The ρ_{ij} now take the place of the π_{ij} in an ordinary ($T_2(1)$) tournament. For example, in (6.4.2) π_{12} has to be replaced by ρ_{12}, π_1 by $\rho_1 = \rho_{1j}$, and π_2 by $\rho_2 = \rho_{2j}$ ($j = 3, \ldots, t$). Handa and Maitri (1984) in fact obtain the probability that a superior player will win $T_2(l, c)$ in the presence of up to two other superior players (the remaining players all being of equal strength). They also give formulae for the expected number of matches, $\mathscr{E}N$, in the case of the single outlier model (6.2.1). In the last case it is therefore possible to find l and c so as to minimize $\mathscr{E}N$, subject to the requirement that the probability of correct selection is at least P' (pre-assigned).

For $t=4$, 5, 7 Handa and Maitri (1984) have compared $T_2(c)$, $T_2(l, c)$, and a repeated RR under the one-outlier model with $\pi=0{\cdot}75(0{\cdot}5)0{\cdot}90$ and with $P'=0{\cdot}75$, $0{\cdot}90$, $0{\cdot}95$. In all cases the expected number of games is a minimum for $T_2(l, c)$, except that in a number of cases $T_2(c)$ is the same as $T_2(l, c)$. The superiority of $T_2(l, c)$ in correctly picking the top player persists when curtailed Round Robins are added to the comparison (David and Andrews, 1987).

Other tournament problems

Many years ago Lewis Carroll (e.g., 1947) expressed dissatisfaction with Knock-out tournaments as used in tennis because of their obvious lack of reliability in picking the second-best player. Carroll supposed that the competitors could be arranged unambiguously in order of strength and that a player would always defeat a weaker opponent. Under these assumptions it is clear that the best player must triumph without fail. To remedy possible injustice to the second-best player, Carroll gave a detailed prescription, the essential feature being that a player, A, is eliminated only when it becomes clear that two others are superior to A. This might be by direct defeat at their hands (as in a double elimination tournament) or indirectly when A suffers a first loss against an opponent already once beaten.

A swifter, if less sporting, way would be to arrange a KO tournament between all the players defeated by the champion. To see how many matches are required for this procedure, consider first an ordinary KO tournament with t players where, quite generally,

$$t=2^p+q \quad (1\leqslant q<2^p).$$

Clearly, q preliminary matches are needed to reduce the number of players to 2^p, so that the tournament is completed in

$$q+(2^p-1)=t-1 \text{ matches.}$$

Since the champion may have defeated as many as $p+1$ opponents, p further matches will determine the runner-up, making the total

$$(6.4.10) \qquad t-1+p=t-1+[\log_2(t-1)],$$

where square brackets denote the integral part here and for the rest of this section.

The result (6.4.10) is given by Steinhaus (1950), who also considers the problem of ensuring a complete ranking in as few matches as possible. He supposes that a set of players has already been ranked and that we have to find a place among them for a new one, A.. We seek first the medium player M, i.e., the one ranked in the middle—if the number of players is even, we call both players in the middle medians. Then we match A with M. If $A \rightarrow M$ we play A against the median of the upper half; if $A \leftarrow M$, against the median of the lower half, and so on, until A's position is determined. Thus since 3 matches are needed to rank 3 players (two *may* do so, but that is not good enough) four can always be ranked in 5 matches, five in 8 matches, etc. To insert the $(k+1)$th player requires $S(k)=1+[\log_2 k]$ matches. Hence the general formula for the number of matches needed to rank t players is

$$\begin{aligned} M(t) &= S(1)+S(2)+\ldots+S(t-1) \\ &= (t-1)+[\log_2 2]+[\log_2 3]+\ldots+[\log_2(t-1)]. \end{aligned}$$

We now express t in the form

$$t=2^{r-1}+s \quad (0 \leqslant s < 2^{r-1}),$$

so that $r=1+[\log_2 t]=S(t)$. Then

$$\begin{aligned} M(t) &= t-1+2(1)+2^2(2)+2^3(3)+\ldots+2^{r-2}(r-2)+s(r-1) \\ &= t-1+2\{1+(r-3)2^{r-2}\}+(t-2^{r-1})(r-1) \\ &= 1+rt-2^r. \end{aligned}$$

Katz (1977) has obtained an expression for the corresponding expected number of matches $E(t)$ (see Table 6.2) and also deals with the problem of correctly ranking just the top $k(\leqslant t)$ players.

An ingenious shorter method, due to Ford and Johnson (1959), has aroused considerable interest in the computing literature because it provides a theoretically very efficient technique for ordering (or "sorting") t numbers or t items that can be ordered linearly. See the interesting account of this and related methods in Knuth (1975, p. 181 ff.). Although the Ford–Johnson algorithm is not optimal (Manacher, 1979), it is close to being so and not difficult to apply.

Suppose $t = 2i$ or $2i + 1$. Then

(a) Pair off $2i$ of the players and let the pairs play in the first round, leaving one competitor out if t is odd.

(b) By a continuation of the present method applied to i players, give a complete ranking of these i first-round winners.

(c) The third step is best explained by a diagram.

At this point in the ranking we have a hierarchy of the form illustrated in Fig. 6.2 for $t = 19$. First-round winners $J, I, \ldots, B$ are ranked in that order, with J the best player. A is the first-round loser to B, and the other first-round losers are placed directly below their respective victors. The odd player with the first-round bye is considered a loser and put in the position at the extreme left of the diagram.

The phrase "main chain" initially will refer to the chain $JIH\ldots CBA$, and the procedure will be to insert the numbered points in the main chain in the order indicated. The procedure is based on the fact that the insertion of a single point into a chain by the Steinhaus method is most efficient if the number of points on the chain is of the form $2^K - 1$.

Hence we start by inserting the point 1 in the chain ABC. After this has been done, the "main chain" under point 2 consists of AB and possibly point 1; this insertion can also be performed with two comparisons.

We now turn to chains of length $2^3 - 1 = 7$, and observe that point 3 is as high as we can go, the main chain under 3 being composed of $ABCDE$ and 1 and 2; and so forth.

This represents a ranking technique, requiring $U(t)$ comparisons, where $U(t)$ is given by the following recurrence relations:

$$U(2k) = k + U(k) + \sum_{j=2}^{k} T(j),$$

$$U(2k+1) = k + U(k) + \sum_{j=2}^{k+1} T(j),$$

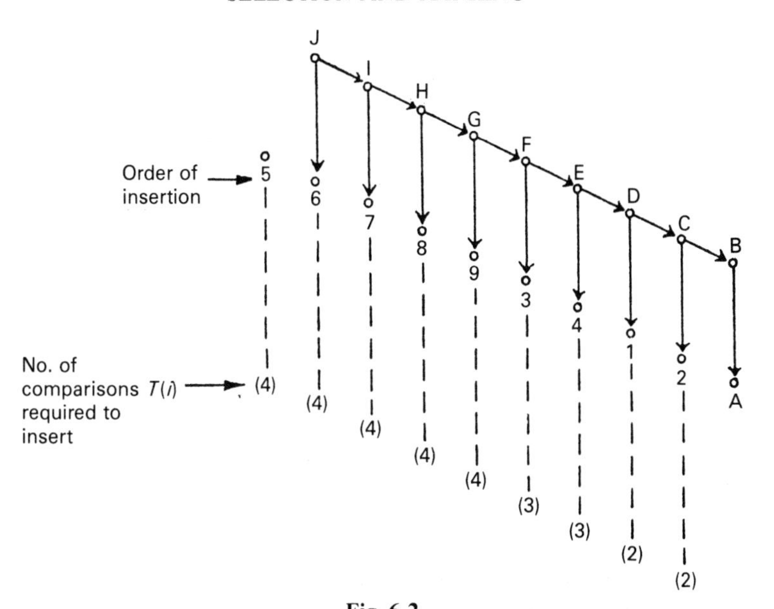

Fig. 6.2

From Ford and Johnson (1959) by permission of the authors and the Editor of the *American Mathematical Monthly*

with $U(1)=0$, $U(2)=1$; the $T(j)$ are the number of comparisons required to insert a player into a chain of length j, and

$$\begin{aligned} T(j) &= 2 \quad \text{for} \quad 1 < j \leqslant 3, \\ &= 3 \quad \text{for} \quad 3 < j \leqslant 5, \\ &= 4 \quad \text{for} \quad 5 < j \leqslant 11, \\ &\dots \\ &= h \quad \text{for} \quad t_{h-1} < j \leqslant t_h, \end{aligned}$$

where $t_h = \frac{1}{3}\{2^{h+1} + (-1)^h\}$.

In Table 6.2 we list values of $M(t)$, $E(t)$, $U(t)$ for $t \leqslant 13$. The last line gives values of $L(t)$, a lower bound which Ford and Johnson obtain from information theory. Each pairing can do no more than divide the remaining possibilities into two complementary sets; the result of the comparison then selects one or the other of these. Observing $t!$ possibilities initially, with halving, the best we can do at each stage, we are led to the formula

$$L(t) = \text{the smallest integer} \geqslant \log_2 t!.$$

Table 6.2

From Ford and Johnson (1959) by permission of the authors and the Editor of the *American Mathematical Monthly*

t	1	2	3	4	5	6	7	8	9	10	11	12	13
$M(t)$	0	1	3	5	8	11	14	17	21	25	29	33	37
$E(t)$	0	1	2·7	4·7	7·1	9·7	12·2	15·6	18·8	22·2	25·8	29·4	33·2
$U(t)$	0	1	3	5	7	10	13	16	19	22	26	30	34
$L(t)$	0	1	3	5	7	10	13	16	19	22	26	29	33

Notes

(1) In a replicated Round Robin tournament of t equal players, suppose that A_i $(i=1,\ldots,t)$ has rank r after m replications. What is A_i's probability of having rank s after $n(>m)$ replications? This question is studied by Yang (1977).

(2) Picking the better of just two players is a problem of interest in its own right. It has received attention in this chapter only as a component part of the larger tournaments $T_2(c)$ and $T_2(l, c)$ of §6.4. A different method of picking the better of two is to require one player to win at least c matches *and* to lead the other by at least l wins. This approach has been studied by George (1974) through its relation to a Wald sequential probability ratio test.

EXERCISES

6.1 Two tennis players A and B have a (friendly) match consisting of n sets. A wins a set from B with constant probability π. If they finish off after an even number of sets $(n=2r)$ with the score at $r{:}r$ they toss for the honours. Show that A's probability of winning the match is the same for $n=2r-1$ as for $n=2r$.

6.2 In a strong Round Robin tournament with $t \geqslant 4$, suppose $a_1=a_2=a$ and $A_1 \rightarrow A_2$. Show that the limiting Kendall–Wei method ranks A_1 ahead of A_2 if $a=t-2$ and A_2 ahead of A_1 if $a=1$.

6.3 Suppose that in a Round Robin of n rounds the row-sum score a_i is replaced by

$$\sum_{j=1}^{t}\left(\frac{\alpha_{ij}}{n(t-1)-a_j}\right)a_j \quad i=1,2,\ldots,t \quad (\alpha_{ii}=0).$$

Interpret this re-allocation and show that if the allocation process is continued, the limiting scores v_i satisfy (6.1.2).

(Daniels, 1969)

6.4 In a Round Robin tournament of four players one of them has probability $\pi>\frac{1}{2}$ of defeating each of the others who are of identical strength. Show that if ties for first place are broken by lot the best player wins the tournament with probability π_2, the same as in a Knock-out tournament.

6.5 The number of possible ways of matching $t=2^p$ players in a KO tournament is

$$\frac{t!}{2^{t-1}}$$

(Narayana and Agyepong, 1980)

6.6 For a KO-tournament with $t=2^p$ players show that for the one-outlier model (6.2.1) the probability $R_t^{(i)}$ that A_t plays exactly i games is given by

$$R_t^{(i)}=\pi^{i-1}\bar{\pi}(i<p), \quad R_t^{(p)}=\pi^{p-1}, \quad (\bar{\pi}=1-\pi).$$

Hence show that the expected number of games played by A_t is $(1-\pi^p)/\bar{\pi}$ and that the probability of A_t and A_1 meeting in the course of the tournament is

$$\frac{1-\pi^p}{\bar{\pi}(t-1)}.$$

[Note that the last result is a special case of (6.4.3).]

(Cf. Narayana, 1979, p. 58)

6.7 For a random KO tournament with t players and k rounds, let t_i be the number of players beginning round i ($t_1=t$), m_i the number of matches in round i, and $p_i=2m_i/t_i=1-q_i$ ($i=1, 2,\ldots,k$) the probability that a specified player will have a match in round i. Let P_i be the probability that A_t wins the tournament having reached round $k-i+1$, and let R_i be the number of matches played by A_t in the last i rounds. Show that under the one-outlier model (6.2.1)

(a) $P_k = (p_1 \pi + q_1) P_{k-1}$

(b) $P_k = (p_1 \pi + q_1) \dots (p_{k-1} \pi + q_{k-1}) \pi$,

(c) $\mathscr{E}(\mathrm{R}_\mathrm{k}) = p_1 \bar{\pi} + p_1 \pi [\mathscr{E}(R_{k-1}) + 1] + q_1 \mathscr{E}(R_{k-1})$,

(d) $P_k + \bar{\pi} \mathscr{E}(R_k) = 1$.

[Hint: Use (a), (c), and induction to prove (d).]

(Narayana, 1979, p. 62)

7 A MISCELLANY

7.1 Selection of judges

In our treatment so far we have paid no attention to the important practical problem of choosing a panel of judges. The choice is, of course, influenced by such matters as experimental cost and availability of suitable judges. Even more basic is the purpose of the paired-comparison experiment for which the judges are to be selected. A large *consumer panel*, representative of the population of interest, is needed to study preferences in the field, while a small *expert panel* is used to detect fine differences in the laboratory and to uphold quality standards.

We shall say more about consumer panels in §7.2. With expert panels primary concern lies in the detection of differences between objects, one or more of which may be "standards". Every attempt should be made to define the kind of difference to be looked for. In the food industry the selection of judges is commonly based on their performance with simple arrangements, such as the pair, triad, and duo–trio tests. Before discussing these we note that they involve only two different types of objects. In order to cover a wider range the experimenter might set up a balanced paired-comparison experiment with several objects. The judges with the fewest circular triads if there is no prior ordering, or with the smallest number of separations (§2.1) plus circular triads if there is a prior ordering, would then be chosen.

Ranking (Ura, 1960b) and scoring (Ferris, 1957) may also be used as selection techniques, but this seems less desirable when the subsequent experimentation is to be by paired comparisons. See also Bock (1956).

Pair, triangle, and duo–trio tests

In a pair test two different objects are presented to a judge. One of these is "superior" on some criterion and a correct response is obtained if the judge identifies the superior object. The other two tests involve three objects, two of which are alike. Call these A_1, A_1', and let A_2 denote the odd object. The judge attempts to pick A_2 in both cases; for the duo–trio test the identity of either A_1 or A_1' is also known. It is seen that the probabilities of correct response purely by chance are $\frac{1}{2}$ for the pair and duo–trio tests, and $\frac{1}{3}$ for the triangle test.

Several properties of these tests have been examined by Ura (1960a) in a fine paper which is, however, at times concise to the point of unkindness. Further details are given by David and Trivedi (1962). Let the registered merits of A_1, A_1', A_2 be respectively y_1, y_1', y_2. Ura assumes in effect that a linear model applies and that the y's are independent continuous variates. Without loss of generality we may take A_2 to be a superior object. Correct responses are obtained if

$$\begin{aligned} & y_1 < y_2 && \text{pair test,} \\ (7.1.1)\quad & |y_1 - y_1'| < |y_1 - y_2| \text{ and } |y_1 - y_1'| < |y_1' - y_2|, && \text{triangle test,} \\ & |y_1 - y_1'| < |y_1 - y_2| && \text{duo–trio test, } A_1 \text{ identified.} \end{aligned}$$

Denote the corresponding probabilities by P_P, P_Δ, and P_D. For y_1, y_1' normal $N(\mu_1, \sigma^2)$, and y_2 normal $N(\mu_2, \sigma^2)$ $(\mu_2 > \mu_1)$ we have

$$P_P = \Phi(\delta/\sqrt{2}),$$

where Φ is the unit normal cdf and $\delta = (\mu_2 - \mu_1)/\sigma$. It can be shown that

$$(7.1.2)\qquad P_\Delta = 2\int_0^\infty [\Phi(\sqrt{\tfrac{2}{3}}\delta - \sqrt{3}x) + \Phi(-\sqrt{\tfrac{2}{3}}\delta - \sqrt{3}x)]\, d\Phi(x),$$

and (Exercise 7.1)

$$(7.1.3)\qquad P_D = 1 - \Phi(\delta/\sqrt{2}) - \Phi(\delta/\sqrt{6}) + 2\Phi(\delta/\sqrt{2})\Phi(\delta/\sqrt{6}).$$

Ura also obtains the corresponding expressions when the y's have uniform distributions with the above means and variances (Exercise 7.2). The results are plotted as functions of δ in Fig. 7.1 and are surprisingly close for the two distributions. Corresponding tables are given by Bradley (1963). See also Frijters (1982) for a detailed table of P_Δ in (7.1.2) and Frijters (1979) for a corresponding expression in the logistic case.

In the triangle test the probability of selecting A_1 is $\frac{1}{2}(1-P_\Delta)$. If, with A_1' identified, we suppose that P_D is simply the probability of choosing A_2 from the reduced set A_1 and A_2[(*)], we have

$$P_D = \frac{P_\Delta}{P_\Delta + \frac{1}{2}(1-P_\Delta)} = \frac{2P_\Delta}{1+P_\Delta}.$$

This simple relation holds to a good approximation in both the normal (Bradley, 1963) and the uniform cases.

If, for the moment, we regard the three arrangements as means of detecting differences between two similar objects (rather than as means of choosing judges) we can compare their performance as tests of the null hypotheses $P_P=\frac{1}{2}$, $P_\Delta=\frac{1}{3}$, $P_D=\frac{1}{2}$ against the respective alternatives $P_P>\frac{1}{2}$, $P_\Delta>\frac{1}{3}$, $P_D>\frac{1}{2}$. Suppose the number of replications is $n=20$. Then the power functions of the three binomial tests can easily be found with the help of the normal approximation to the binomial. The advantage lies overwhelmingly with the pair test. Thus while, at a 5 per cent level of significance, a difference $\mu_2-\mu_1=1{\cdot}5\,\sigma$ is detected with probability 0·95 by the pair test, the corresponding probabilities with the other tests are less than $\frac{1}{2}$. The superiority of the pair test over the duo–trio test is apparent from Fig. 7.1, but the triangle test is only a little better than the duo–trio. These theoretical results are in line with experimental investigations made by Byer and Abrams (1953), Hopkins and Gridgeman (1955), as well as Ura.

When selecting judges we are not so much interested in whether they have any ability at all, but rather in whether their ability is sufficiently great. We may characterize each judge's discriminating power by a personal σ-value and choose only those judges for whom $\sigma \leqslant \sigma_0$, where σ_0 corresponds to a prescribed degree of ability. For a difference $\Delta\mu=\mu_2-\mu_1$ between A_2 and A_1 the probability of a correct response may be expressed as $p(\Delta\mu/\sigma)$.[(†)] Since this is a decreasing function of σ, for $\Delta\mu$ fixed, the hypotheses

(7.1.4) $\quad H_0\colon \sigma \leqslant \sigma_0 \quad \text{and} \quad H_1\colon \sigma > \sigma_0$

(*) This is just the Luce postulate (§1.4).

(†) This probability also depends on which of the three arrangements is used, but we omit the subscripts P, Δ, D for convenience.

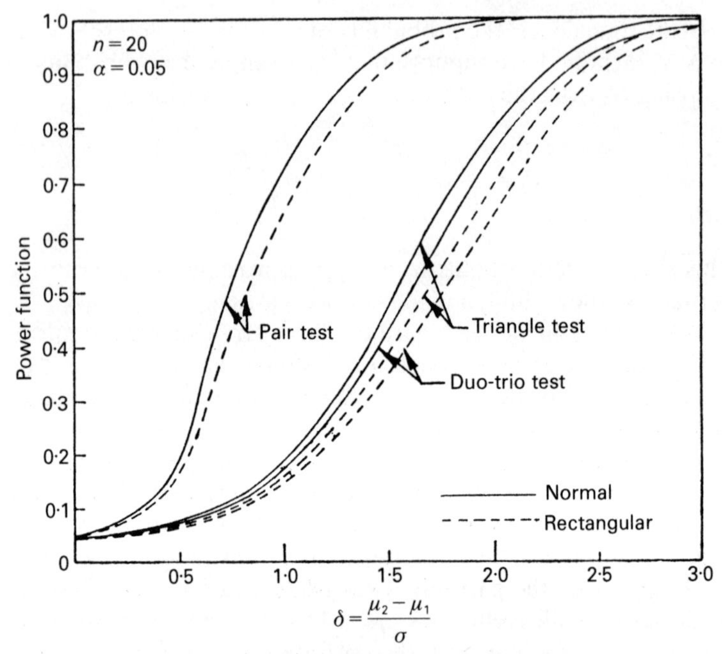

Fig. 7.1 Probabilities of correct response

From Ura (1960a) by permission of the author and the Editor of *Reports of Statistical Application Research,* Union of Japanese Scientists and Engineers

are equivalent to

(7.1.5) $H_0 : p(\Delta\mu/\sigma) \geqslant p(\Delta\mu/\sigma_0)$ and $H_1 : p(\Delta\mu/\sigma) < p(\Delta\mu/\sigma_0)$.

When a judge scores x correct responses in n trials we reject H_0, applying the arcsine transformation, if

$$\sin^{-1}\surd(x/n) \leqslant \sin^{-1}\surd p(\Delta\mu/\sigma_0) - u_\alpha/\surd(4n),$$

where u_α is the upper α significance point of a unit normal variate. The power function of this test is approximately

$$\Phi\{\surd(4n)[\sin^{-1}\surd p(\Delta\mu/\sigma_0) - \sin^{-1}\surd p(\Delta\mu/\sigma)] - u_\alpha\},$$

a function of $\Delta\mu/\sigma_0$ as well as σ. Since the size of $\Delta\mu$ can be varied by the experimenter the question arises as to what its optimal value is. Ura concludes from a study of power functions that, for $\alpha = 0{\cdot}05$, $\Delta\mu$ should be approximately such that the probability of a correct response is 0·95 for a judge who is just admissible ($\sigma = \sigma_0$). Moreover, he finds the triangle test best in these circumstances. For an attempt to improve the triangle test, by asking judges to score the degree of difference between the selected object and the other two, see Bradley and Harmon (1964). Regarding such modifications of the triangle test see also Gridgeman (1970) and Frijters (1979).

Since (7.1.5) states simply a null and alternative hypothesis on a binomial parameter, sequential methods for testing a binomial proportion may be expected to reduce, on the average, the number of replications to be made by each judge (Bradley, 1953). In addition to σ_0 we now require a value $\sigma_1 > \sigma_0$ such that judges with $\sigma \geqslant \sigma_1$ are unacceptable. We then test sequentially

$$H_0: p(\Delta\mu/\sigma) = p(\Delta\mu/\sigma_0) \quad \text{against} \quad H_1: p(\Delta\mu/\sigma) = p(\Delta\mu/\sigma_1).$$

Alternatively, one can work directly with the probabilities p of correct response, fix suitable values p_0 and p_1 ($< p_0$) of p, and test

$$H_0: p = p_0 \quad \text{against} \quad H_1: p = p_1.$$

This is at the same time a test of

$$H_0: p \geqslant p_0 \quad \text{against} \quad H_1: p \leqslant p_1.$$

It may also be noted that in the absence of any clear guidance for the determination of acceptable values of σ or p, multiple decision procedures on binomial populations (Gupta and Sobel, 1960) can be used to screen out inferior judges.

The conditions (7.1.1) for a correct response in triangle and duo–trio tests continue to apply if the merit of an object requires a k-dimensional representation ($k \geqslant 1$). Correspondingly y_1, y_1', y_2 must be interpreted as k-dimensional vectors and, e.g., $|y - y_1'|$ as the Euclidean distance between y_1 and y_1'. Ennis and Mullen (1986a, b) have investigated the triangle test in some detail, using various assumptions of multivariate normality of the vectors y_1, y_1', y_2. They point out that even if A_1 and A_2 *differ* in merit in only one dimension, the probability of correct selection will be lower than given by P_Δ if other dimensions enter the judgment process.

Additional information on many of the topics of this section is given in Gacula and Singh (1984, Chapter 10).

7.2 Consumer tests

The selection of a consumer panel presents the familiar difficulties of finding a representative and yet co-operative group. Once chosen, members of the panel should be confronted with simple, well-defined alternatives. Paired comparisons requiring a decision in favour of one of two alternatives (and possibly permitting also a declaration of no preference) will often provide the natural procedure, although it may be possible to elicit supplementary information from the respondents. The experimenter must attempt to balance out extraneous effects, such as the order of presentation. Precautions should be taken to disguise any identifying marks on the objects under comparison if, for example, a taste preference is required. Sometimes only a single pair can be judged by one panel member as the work involved may be considerable, e.g. in the comparison of two recipes. Since experiments with large consumer panels are liable to end up incomplete it is wise not to attempt too much. In so far as interest usually centres on obtaining from the panel a general picture of where their preferences lie, linear models are not likely to apply; if they do, then methods for dealing with incomplete data mentioned in Chapter 4 will be useful. Finally, in this list of rather obvious but important points, we note that it would be inappropriate to instruct members of a consumer panel on what they should base their preferences: these are wanted in their natural state.

Models for consumer testing

Individuals with poor discriminating powers have no place on an expert panel. However, in a consumer test it is of decided interest to know something about the proportion of subjects who cannot distinguish between two products A and B. Such knowledge is of value in helping one decide on whether to market A or B or both. In situations where the undiscriminating panel members can be identified the comparison of A and B will be sharpened by their omission (cf. Bliss, 1960).

It may be supposed, as a first approximation to the true situation (Ferris, 1958), that a proportion π_0 of the population under study is unable to distinguish A and B, and that π_a, π_b are the proportions preferring A, B $(\pi_a + \pi_o + \pi_b = 1)$. Let $P_i(j)$ be the probability that a consumer of type $i(i = a, o, b)$ prefers product j

on any particular occasion when A and B are presented. We may take j to be A, B, or O, where O denotes no preference. Ferris assumes that

$$P_a(A) = P_b(B) = 1,$$

$$P_a(B) = P_b(A) = P_a(O) = P_b(O) = 0,$$

$$P_o(O) = 1 - 2p,$$

and

$$P_o(A) = P_o(B) = p, \quad \text{where} \quad 0 \leqslant p \leqslant \tfrac{1}{2}.$$

This means that those who really prefer A (or B) will consistently do so. Others may be unable to discriminate or, while able to discriminate, be inconsistent in their preferences; in either case they prefer A (or B) with probability p, and have no preference with probability $1 - 2p$.

If n consumers are asked to judge A and B twice (without their knowledge that the same pair is before them on both occasions) it is possible to estimate all the parameters (π_a, π_o, π_b, p) by the method of maximum likelihood.

Approximate simultaneous confidence regions for π_a, π_b, π_o, based on asymptotic theory, are obtained by Kalmijn (1976). Modelling of such repeat paired preference (RPP) testing is elaborated in Horsnell (1969) where many general issues are also discussed. For example, it is not at all clear how to translate the results of consumer tests into actual longterm consumer behaviour, especially if A and B differ in price or other characteristics not part of the preference judgment (e.g., butter and margarine).

Questions of the practical adequacy of Ferris's model and its extensions are debated by Wierenga (1974), Horsnell (1977), and Hutchinson (1979). One issue is the proper handling of responses of no preference (ties) between A and B.

7.3 Treatment of ties

There is an almost embarrassing number of ways of dealing with ties in individual comparisons. We can prevent them by brute force, instructing an undecided judge to toss a metal coin in favour of one of the two objects; or, allowing the judge the comfort of reserving judgment, we can let a physical coin decide instead. These two methods have the advantage that the resultant data can be analysed by the methods of the foregoing chapters. The same is not strictly true if, for each pair, the ties are simply divided equally. Nevertheless, this method has the advantage of not depending on more or less arbitrary ways of breaking ties; in particular, physical randomization can lead to different conclusions with the same experimental results. Another possibility is to do the randomization theoretically. Consider for example, the coefficient of agreement u of (2.1.5) (Kendall, 1962, §11.8; van der Heiden, 1952):

$$u = \frac{2\sum_{i \neq j}^{t} \binom{\alpha_{ij}}{2}}{\binom{n}{2}\binom{t}{2}} - 1. \tag{7.3.1}$$

If the n comparisons of A_i and A_j result in a ties, equal division of these would lead us to substitute $x + \frac{1}{2}a$ for α_{ij} and $n - x - \frac{1}{2}a$ for α_{ji}, where x is the number of clear preferences for A_i. A theoretical randomization is achieved by replacing the terms $\binom{\alpha_{ij}}{2} + \binom{\alpha_{ji}}{2}$ of (7.3.1) by

$$\sum_{s=0}^{a} \binom{a}{s} 2^{-a} \left\{ \binom{x+s}{2} + \binom{n-x-s}{2} \right\}, \tag{7.3.2}$$

that is, by considering all outcomes to which the random breaking of the a ties can give rise, and weighting them by their respective probabilities of occurrence. The expression (7.3.2) may be shown to exceed

$$\left\{ \binom{x + \frac{1}{2}a}{2} + \binom{n - x - \frac{1}{2}a}{2} \right\},$$

the value under equal division of ties, by $\frac{1}{4}a$. Correspondingly, u is easily calculated under a theoretical randomization.

For $t = 2$, Draper, Hunter, and Tierney (1969) have proposed a Bayesian approach. In the comparison of A and B, let x, y be the respective number of clear preferences and π_A, π_B the corresponding preference probabilities. On the assumption of a non-informative (flat) Dirichlet prior of π_A, π_B, the posterior density of π_A and π_B is

$$\frac{(n+2)!}{x!\,y!(n-x-y)!}\,\pi_A^x\,\pi_B^y(1-\pi_A-\pi_B)^{n-x-y} \qquad \pi_A+\pi_B \leqslant 1.$$

Contours of this distribution can be plotted and used to make inferences.

In an appraisal of these and other procedures for handling ties it is helpful to distinguish between hypothesis testing and estimation. For $t=2$, when the method of paired comparisons reduces to a sign test, it has been shown that ignoring ties altogether leads to a more powerful test than either equal or random division of ties (Hemelrijk, 1952; Putter, 1955). This interesting result, which does not extend to $t>2$, applies to the analysis of data containing ties; a related question is whether ties should be permitted in the first place. It would seem good strategy to instruct judges to express a preference only when they are reasonably sure. The experimental results are then undiluted by guesses. However, the situation is not clear-cut in practice as the admission of ties may make the judge more lax (Gridgeman, 1959).

When we turn to estimation it is apparent that ties included in the experimental results should not be ignored: with the same numbers of clear preferences, two objects A_i, A_j are closer in merit the more ties there are between them. If the clear preferences favour A_i it may be argued that A_i is likely to have had "the better of the ties". This runs counter to all the above "equitable" methods of treating ties but may be achieved by the following model (Glenn and David, 1960). As usual, let y_i, y_j be the observed merits of A_i, A_j. Suppose that the judge declares a tie if $|y_i-y_j| \leqslant \tau$ $(\tau \geqslant 0)$, prefers A_i if $y_i-y_j>\tau$, and prefers A_j if $y_j-y_i>\tau$. The model can be easily superimposed on that of Thurstone and Mosteller, and on the inverse sine law of (4.2.2). Least squares methods can then be used to estimate not only the merits V_i $(i=1, 2,\ldots,t)$ but also the threshold parameter τ (and possibly different τ's for different judges, a point which can be tested). No splitting of the ties is actually made, the original observations being used in the analysis. See also Greenberg (1965).

Bradley–Terry model

Rather simple expressions can be obtained for the probabilities of the three possible outcomes if the foregoing threshold model is superimposed on that of Bradley and Terry (Rao and Kupper, 1967). From (1.3.10) we have that for $i, j = 1, 2, \ldots, t, i \neq j$,

$$\pi_{ij} = \tfrac{1}{4} \int_{-(V_i - V_j) + \tau}^{\infty} \operatorname{sech}^2 \tfrac{1}{2} y \, dy$$

$$= \frac{\pi_i}{\pi_i + \theta \pi_j}, \tag{7.3.3}$$

where $\theta = e^{\tau}$. It follows that the probability, ψ_{ij}, of a tie is

$$\psi_{ij} = 1 - \pi_{ij} - \pi_{ji}$$

$$= \frac{\pi_i \pi_j (\theta^2 - 1)}{(\pi_i + \theta \pi_j)(\theta \pi_i + \pi_j)}.$$

Likelihood methods can now be applied for estimation and hypothesis testing by generalizations of the procedures for the case without ties ($\theta = 1$).

A different approach for handling ties under the Bradley–Terry model has been put forward by Davidson (1970, 1973). He supposes that

$$\psi_{ij} = \nu \sqrt{\pi_{ij} \pi_{ji}}, \tag{7.3.4}$$

where ν^{-1} (>0) may be regarded as a discrimination constant. This choice gives

$$\pi_{ij} = \frac{\pi_i}{\pi_i + \pi_j + \nu \sqrt{\pi_i \pi_j}}, \tag{7.3.5}$$

which preserves the relation $\pi_{ij}/\pi_{ji} = \pi_i/\pi_j$ entailed by Luce's (1959) axiom of choice. Under what conditions this relation remains appropriate in the presence of ties is another matter. For the threshold model we see from (7.3.3) that $\pi_{ij}/\pi_{ji} > \pi_i/\pi_j$ for $\theta > 1$ and $\pi_i > \pi_j$, which reflects the notion that preferences for the untied pairs tend to be more sharply divided than if no ties are allowed.

Davidson (1970) also obtains the interesting result that the ML estimates of the π_i in a balanced paired-comparison experiment with ties rank the objects according to their scores (in which a tie is counted as half a win). This generalization of Zermelo's (1929) and Ford's (1957) result in the absence of ties (4.3) does not hold in general for the threshold model for which the scores are no longer sufficient statistics.

Implementation of both the Rao–Kupper and Davidson models naturally requires iterative methods since the case of no ties does. Such methods are not needed for the locally asymptotically most stringent tests proposed by Beaver (1974) for each of the two models. He shows that all four procedures have the same asymptotic relative efficiency for testing the hypothesis of no preference ($\pi_1 = \pi_2 = \ldots = \pi_t$). See also Singh and Gupta (1975, 1978) who arrive at and examine the model

$$\pi_{ij} = \pi_i \eta_{ij}, \; \psi_{ij} = \pi'_{ij} \eta_{ij},$$

where

$$\pi'_{ij} = \pi'_{ji} \geqslant 0 \quad \text{and} \quad \eta_{ij} = (\pi_i + \pi_j + \pi'_{ij})^{-1}.$$

Davidson's model (7.3.4–5) results if $\pi'_{ij} = \nu\sqrt{\pi_i \pi_j}$.

Both the Rao–Kupper and the Davidson methods have been extended by Beaver and Rao (1973) to handle ties in the Bradley–Pendergrass triple comparison model (1.4.2). Three models for multivariate paired comparisons with ties have been proposed by Davoodzadeh and Beaver (1982).

Models allowing for ties and individual judge effects—also order of presentation effects which we shall not consider here—have been proposed by Kousgaard (1976). Let $x_{ij\gamma} = -1, 0, 1$ according as, for the γth judge, $A_i \leftarrow A_j$, A_i ties A_j, $A_i \rightarrow A_j$, with respective probabilities

$$\pi_{ji\gamma}, \; \psi_{ij\gamma}, \; \pi_{ij\gamma} \; (1 \leqslant i < j \leqslant t, \; 1 \leqslant \gamma \leqslant n).$$

Then Kousgaard's model IV is

$$\Pr\{x_{ij\gamma} = x\} = \exp[\theta_{ij} x + \xi_\gamma (1 + x)(1 - x)] / \zeta_{ij\gamma}, \tag{7.3.6}$$

where

$$\zeta_{ij\gamma} = e^{\theta_{ij}} + e^{-\theta_{ij}} + e^{\xi_\gamma}$$

and the θ_{ij}, ξ_γ can range from $-\infty$ to ∞.

We see that $\theta_{ij} = \frac{1}{2}\log(\pi_{ij\gamma}/\pi_{ji\gamma})$. If $\theta_{ij} = \theta_i - \theta_j$ and $\xi_\gamma = \log \nu$ for $\gamma = 1, 2, \ldots, n$, we get Davidson's model (7.3.4–5) on setting $\theta_i = \frac{1}{2}\log \pi_i$, $i = 1, 2, \ldots, t$.

The θ_{ij} may be estimated by maximizing the likelihood function corresponding to (7.3.6), conditioned on $u_1, \ldots, u_n$, where $u_\gamma = \Sigma_{i<j}(1+x_{ij\gamma})(1-x_{ij\gamma})$, the number of ties declared by judge γ. See Kousgaard (1976) for the rather complicated procedure.

Kousgaard (1984) has shown how a model of the type (7.3.6) can be modified to incorporate concomitant variables, say $\mathbf{z}_i' = (z_{i1}, \ldots, z_{ip})$ for the ith object. Dropping the judge effect, he uses standard methods on the model

$$\Pr\{x_{ijk} = x\} = \exp[(\theta_i - \theta_j)x + \xi(1+x)(1-x)]/\zeta_{ij},$$

where $\theta_i = \boldsymbol{\beta}'\mathbf{z}_i$, to study the effects of the concomitant variables. In an interesting application to acoustics there is the further feature that the θ_i have a factorial structure.

Weak stochastic ranking

In the context of weak stochastic ranking (§2.2) Singh and Thompson (1968) were the first to incorporate ties into paired-comparison experiments. They give an interesting discussion of the structure of such experiments in terms of *bigraphs*, i.e., graphs consisting of both directed arcs (for wins and losses) and undirected arcs (for ties).

If ψ_{ij} is the probability of a tie when A_i meets A_j, then in generalization of (2.2.1) the likelihood function is

$$L = \prod_{i<j} \binom{n_{ij}}{\alpha_{ij}, \beta_{ij}, \alpha_{ji}} \pi_{ij}^{\alpha_{ij}} \psi_{ij}^{\beta_{ij}} \pi_{ji}^{\alpha_{ji}}, \tag{7.3.7}$$

where

$$\alpha_{ij} + \beta_{ij} + \alpha_{ji} = n_{ij}, \quad \pi_{ij} + \psi_{ij} + \pi_{ji} = 1 \quad (1 \leqslant i < j \leqslant t).$$

It is rather natural to proceed by maximizing L over the space (deCani, 1969)(*)

$$S_{ij} = (\pi_{ij}, \psi_{ij} \colon \psi_{ij} \geqslant 0, \quad \pi_{ij} \geqslant 1 - \pi_{ij} - \psi_{ij} \geqslant 0).$$

Then, given that A_i precedes A_j in the ranking, the maximization of (7.3.7) is a nonlinear programming problem leading to the intuitively appealing ML estimators

(*) Singh and Thompson use a different space.

$$\hat{\psi}_{ij}=\frac{\beta_{ij}}{n_{ij}},\quad \hat{\pi}_{ij}=\max\left(\frac{\alpha_{ij}}{n_{ij}},\quad \tfrac{1}{2}\frac{(\alpha_{ij}+\alpha_{ji})}{n_{ij}}\right)$$

$$(n_{ij}\geqslant 1,\ 1\leqslant i<j\leqslant t).$$

Taking $n_{ij}=n$, for all i, j, deCani goes on to show how a nearest adjoining order can be obtained by linear programming.

Ties are not really a major complication in weak stochastic ranking. They are specifically allowed for in the general approach of Flueck and Korsh (1975) and in their (1974) branch search algorithm.

7.4 Within-pair order effects

Although Fechner had already regarded "space" and "time" effects as important they have not so far been taken into account in any analysis. Acknowledging that order effects may bias results, we have attempted to balance them out at the design stage (Chapter 5). Unless the effects are large or of interest in themselves this appears to be adequate. Consider, for example, the case of an even number n of comparisons of A and B, $\frac{1}{2}n$ in each order. Let $\pi+\delta$ and $\pi-\delta$ denote the probabilities with which $A\rightarrow B$ when A is presented first and second, respectively, and let a', a'' be the corresponding number of preferences for A. Then $p=(a'+a'')/n$ is an unbiased estimate of the average preference probability π. The variance of p is

$$\frac{1}{n^2}\cdot\tfrac{1}{2}n[(\pi+\delta)(1-\pi-\delta)+(\pi-\delta)(1-\pi+\delta)]=\frac{1}{n}[\pi(1-\pi)-\delta^2],$$

exhibiting the usual reduction due to heterogeneity. It is seen that δ has to be quite sizable before the reduction becomes important. Moreover, the effect of ignoring heterogeneity is to increase the difficulty of establishing a difference between A and B (cf. §3.2).

To deal with situations in which the effect of the order of presentation is of intrinsic interest or is not negligible in the estimation of treatment effects, Beaver and Gokhale (1975) have proposed the following modification of the Bradley–Terry model. With $k=i$ or j, let $\pi_{ij.k}$ denote the probability that $A_i\rightarrow A_j$ when A_k is presented first (i, $j=1, 2,\ldots,t$; $i\neq j$). Then, ties not being permitted, $\pi_{ij.i}$ and $\pi_{ij.j}$ are given by

$$(7.4.1) \qquad \pi_{ij.i} = \frac{\pi_i + \delta_{ij}}{\pi_i + \pi_j}, \quad \pi_{ij.j} = \frac{\pi_i - \delta_{ij}}{\pi_i + \pi_j},$$

where $|\delta_{ij}| \leqslant \min(\pi_i, \pi_j)$.

Since in the B–T model the merit of A_i is measured by $\log \pi_i$ rather than π_i, it is more natural to replace δ_{ij} by a multiplicative parameter γ_{ij} (Davidson and Beaver, 1977):

$$(7.4.2) \qquad \pi_{ij.i} = \frac{\pi_i}{\pi_i + \gamma_{ij}\pi_j}, \quad \pi_{ij.j} = \frac{\gamma_{ij}\pi_i}{\gamma_{ij}\pi_i + \pi_j},$$

where γ_{ij} is attached to the parameter of the object presented second. In contrast to the δ_{ij}, the γ_{ij} are restricted only by $\gamma_{ij} \geqslant 0$. The first position is advantageous or disadvantageous for A_i according as $\gamma_{ij} < 1$ or > 1.

Model (7.4.2) is greatly simplified if $\gamma_{ij} = \gamma$, all i, j $(i \neq j)$, whereas $\delta_{ij} = \delta$, all i, j, implies the restrictive and parameter-dependent inequality $\delta \leqslant \min(\pi_1, \pi_2, \ldots, \pi_t)$.

It is also easy to combine the multiplicative model with the Rao–Kupper and Davidson models for ties (§7.3). For example, with $\gamma_{ij} = \gamma$, (7.3.5) generalizes to

$$\pi_{ij.i} = \frac{\pi_i}{\pi_i + \gamma\pi_j + \nu\sqrt{\pi_i\pi_j}},$$

which implies

$$\pi_{ji.i} = \frac{\gamma\pi_j}{\pi_i + \gamma\pi_j + \nu\sqrt{\pi_i\pi_j}}.$$

Davidson and Beaver (1977) show how maximum likelihood or weighted least squares methods (see §4.6) may be used to estimate the parameters of these models and to test the presence of an order effect, equality of preference, and goodness of fit. Fienberg (1979) and Koehler and Ridpath (1982) handle the same problems by a loglinear model approach for categorical data. Some additional models for ties in the presence of order effects are investigated by van Baaren (1978). See also Kousgaard (1979).

Sirotnik and Beaver (1984) propose a two-stage approach. Initially only nt comparisons of identical items are run. If this shows that order effects are important, the remaining $nt(t-1)$ ordered comparisons are made. The combined nt^2 observations are then analysed according to the Davidson and Beaver (1977) model, thus allowing also for ties.

Within-pair order effects have been combined with the Thurstone–Mosteller model by Harris (1957) and Sadasivan (1983).

7.5 Scheffé's procedure

Scheffé (1952) was actually the first to put forward a method of analysing paired-comparison experiments that made specific allowance for possible within-pair order effects. In addition, he provided a means of analysing data based on a 7- or 9-point preference scale.(*)

In Scheffé's procedure the ordered pairs (A_i, A_j) and (A_j, A_i) must be distinguished. Let s_{ijk} $(k=1, 2,\dots, r)$ be the score for A_i in the kth comparison of the first pair. We use "score" here in a slightly modified sense: $s_{ijk}=3, 2, 1, 0, -1, -2, -3$ for a 7-point scale. Then $-s_{ijk}$ is the score for A_j, whereas s_{jik} is A_j's score in the kth comparison (A_j, A_i). The total score for A_i in all its comparisons with A_j is therefore

$$\sum_{k=1}^{r} (s_{ijk} - s_{jik}).$$

It is assumed that all the s_{ijk} are independent and that for a fixed ordered pair (A_i, A_j) the r variates s_{ijk} have the same mean μ_{ij} and the same variance σ^2 which does not depend on (A_i, A_j). For tests of significance and confidence statements the assumption of normality of scores has to be added. This last assumption can be only approximately true at best, and that of a common variance is also very dubious. Nevertheless it may be hoped that these failures of assumptions will not throw off completely the subsequent analysis of variance, in view of the general robustness of such an analysis for balanced experiments. It is also supposed that each comparison is made by a different judge, requiring $rt(t-1)$ judges in all, and that these are drawn at random from the population of

(*) For comments on such a scale see p. 2.

interest. As we pointed out in §7.1, such an experimental arrangement is applicable to consumer testing but might be modified to cover other situations.

Let the mean preference score for A_i in the ordered pairs (A_i, A_j), (A_j, A_i) be μ_{ij}, $-\mu_{ji}$, respectively. Denote the average of these two means by τ_{ij},

$$\tau_{ij} = \tfrac{1}{2}(\mu_{ij} - \mu_{ji}), \tag{7.5.1}$$

and their difference by $2\delta_{ij}$,

$$\delta_{ij} = \tfrac{1}{2}(\mu_{ij} + \mu_{ji}). \tag{7.5.2}$$

Thus $2\delta_{ij}$ is the difference due to order of presentation in the mean preference for A_i over A_j, while τ_{ij} is the average preference for A_i over A_j, averaged over the two orders. We note

$$\tau_{ji} = -\tau_{ij}, \quad \delta_{ji} = \delta_{ij}.$$

We may also define the average order effect δ by

$$\tfrac{1}{2}t(t-1)\,\delta = \sum_{i<j} \delta_{ij}. \tag{7.5.3}$$

Scheffé now introduces the hypothesis of "subtractivity", which is similar to that of linearity, namely that there exist parameters α_1, $\alpha_2, \ldots, \alpha_t$, for which

$$\tau_{ij} = \alpha_i - \alpha_j \quad \text{for all } i, j \ (i \neq j). \tag{7.5.4}$$

Departures from subtractivity are measured by parameters γ_{ij}, so that in general we may write

$$\tau_{ij} = \alpha_i - \alpha_j + \gamma_{ij}, \tag{7.5.5}$$

where the γ_{ij} are seen to satisfy the conditions

$$\gamma_{ji} = -\gamma_{ij} \quad \text{and} \quad \sum_j{}' \gamma_{ij} = 0.$$

Conversely, the α_i can be expressed in terms of the τ_{ij} by

$$\alpha_i = \sum_j{}' \tau_{ij}/t. \tag{7.5.6}$$

Least-squares estimates of the parameters are immediate. Starting with

$$\hat{\mu}_{ij} = \sum_k s_{ijk}/r,$$

τ_{ij}, δ_{ij}, δ, and α_i can be estimated in turn from (7.5.1), (7.5.2), (7.5.3) and (7.5.6).

The score s_{ijk} may be represented as

$$s_{ijk} = (\alpha_i - \alpha_j) + \gamma_{ij} + \delta + (\delta_{ij} - \delta) + e_{ijk}, \tag{7.5.7}$$

where the error term e_{ijk} is the only random variable on the right-hand side. An unbiased estimate of the variance σ^2 is

$$\hat{\sigma}^2 = S_e/[t(t-1)(r-1)].$$

The error sum of squares S_e is defined in Table 7.1 which gives the split-up of the total sum of squares of scores S_T. Approximate F-ratio tests can be made as indicated. The non-centrality parameters in the last column are defined (Ura, 1957) by

$$\sigma_\alpha^2 = \sum \alpha_i^2/\nu_\alpha, \quad \sigma_\gamma^2 = \sum_{i \neq j} \gamma_{ij}^2/\nu_\gamma, \quad \sigma_\tau^2 = \sum_{i \neq j} \tau_{ij}^2/\nu_\tau,$$

$$\sigma_\delta^2 = \sum_{i \neq j} \delta_{ij}^2/\nu_\delta, \quad \text{and} \quad \sigma_\mu^2 = \sum_{i \neq j} \mu_{ij}^2/\nu_\mu.$$

Scheffé illustrates the analysis and also discusses simultaneous confidence statements for the differences $\alpha_i - \alpha_j$.

A Thurstonian model for Scheffé's analysis of variance has been given by Ura with the help of the following assumptions:

(a) The observed merit y_i $(i = 1, 2, \ldots, t)$ of object A_i is normal $N(V_i, \sigma^2)$.

(b) The random variables y_{ij} representing the preference of A_i over A_j when presented in the order (A_i, A_j) are normal $N(V_{ij}, 2(1-\rho)\sigma^2)$. The V_{ij} may be decomposed:

$$V_{ij} = (V_i - V_j) + \gamma'_{ij} + \delta'_{ij},$$

where the γ'_{ij} are deviations from subtractivity and the δ'_{ij} are order effects, satisfying the same conditions as the γ_{ij}, δ_{ij} above.

(c) For a $(2m+1)$-point scoring system the scale of y_{ij} is divided into $2m+1$ parts, and scores are allotted as follows for $m = 3$:

Table 7.1 Partitioning of the sum of squares of scores

Source	Sum of squares	Degrees of freedom	Expected mean squares
Main effects	$S_\alpha = 2rt \sum_{i=1}^{t} \hat{\alpha}_i^2$	$\nu_\alpha = t - 1$	$\sigma^2 + 2rt\sigma_\alpha^2$
Deviations from subtractivity	$S_\gamma = S_\tau - S_\alpha$	$\nu_\gamma = \binom{t}{2} - (t-1)$	$\sigma^2 + r\sigma_\gamma^2$
Average preferences	$S_\tau = r \sum_{i \neq j}^{t} \hat{\tau}_{ij}^2$	$\nu_\tau = \binom{t}{2}$	$\sigma^2 + r\sigma_\tau^2$
Order effects	$S_\delta = S_\mu - S_\tau$	$\nu_\delta = \binom{t}{2}$	$\sigma^2 + r\sigma_\delta^2$
Means	$S_\mu = r \sum_{i,j} \hat{\mu}_{ij}^2$	$\nu_\mu = 2\binom{t}{2}$	$\sigma^2 + r\sigma_\mu^2$
Error	$S_e = S_T - S_\mu$	$\nu_e = t(t-1)(r-1)$	σ^2
Total	$S_T = \sum_{i,j,k} s_{ijk}^2$	$\nu_T = rt(t-1)$	

−3 −2 −1 0 1 2 3

$-t_3$ $-t_2$ $-t_1$ t_1 t_2 t_3 $y_{ij} \rightarrow$

The probability $P_{ij}(s)$ that A_i will score s against A_j in (A_i, A_j) is consequently

$$P_{ij}(s) = \Phi\left(\frac{t_s - V_{ij}}{\sqrt{[2(1-\rho)]}\,\sigma}\right) - \Phi\left(\frac{t_{s-1} - V_{ij}}{\sqrt{[2(1-\rho)]}\,\sigma}\right),$$

where $\Phi(x)$ is the unit normal cdf. If

$$\max|V_{ij}| \ll \sqrt{[2(1-\rho)]}\,\sigma,$$

that is, for small differences between objects, approximate relations between Ura's and Scheffé's parameters can be found.

Ura also considers the power of Scheffé's test for main effects and uses this in a first attempt to arrive at an optimal scoring system.

Factorial experimentation in the framework of Scheffé's analysis has been treated by Dykstra (1958).

An analysis similar to that in Table 7.1 has been given by Bechtel (1967) in the unreplicated case ($r=1$). This is based on the assumption of equal δ_{ij} in (7.5.7), i.e., on the model

$$s_{ij} = \alpha_i - \alpha_j + \gamma_{ij} + \delta + e_{ij}.$$

Tutz (1986) has extended the threshold approach for paired comparisons with ties to cover also ordered categories.

7.6 Optimal scaling

An interesting method of quantifying paired-comparison data which postulates no specific model has been proposed by Guttman (1946). As usual, let $a_{i\gamma}$ be judge γ's score for A_i, and let $a'_{i\gamma} = t - 1 - a_{i\gamma}$. Further, let $t_\gamma(u_\gamma)$ be the mean of the scores $v_i(\Sigma\, v_i = 0)$ of the objects judge γ ranked higher (lower) than the other objects, weighted by the respective relative frequencies of the judgments:

$$t_\gamma = \sum_i a_{i\gamma} v_i / F, \quad u_\gamma = \sum_i a'_{i\gamma} v_i / F, \tag{7.6.1}$$

where $F = \frac{1}{2}t(t-1)$. Also, let

$$y_\gamma = \sum_i (v_i - t_\gamma)^2 a_{i\gamma} = \sum v_i^2 a_{i\gamma} - t_\gamma^2 F. \tag{7.6.2}$$

$$z_\gamma = \sum_i (v_i - u_\gamma)^2 a'_{i\gamma} = \sum v_i^2 a'_{i\gamma} - u_\gamma^2 F. \tag{7.6.3}$$

The sum of squares of deviations of the scores

$$W = n(t-1) \sum v_i^2 \tag{7.6.4}$$

can be partitioned into two parts

$$R = \sum_\gamma (t_\gamma^2 + u_\gamma^2) F, \tag{7.6.5}$$

and

$$S = \sum_{\gamma} (y_\gamma + z_\gamma), \tag{7.6.6}$$

the sums of squares between and within objects, respectively.

Guttman's approach is akin to discriminant function analysis and consists in finding scores that will minimize the variation within objects compared to that of the group of objects as a whole. This means making S as small as possible or R as large as possible compared with W. If we define the correlation ratio E by

$$E^2 = R/W$$

the required scores v_i are obtained on setting $\partial E^2/\partial v_j$ equal to zero, that is, from the stationary equations

$$\frac{\partial R}{\partial v_j} = E^2 \frac{\partial W}{\partial v_j}. \tag{7.6.7}$$

The derivatives of R can be evaluated from (7.6.5) and (7.6.1) as

$$\frac{\partial R}{\partial v_j} = \frac{2}{F} \sum_i v_i \sum_{\gamma} (a_{i\gamma} a_{j\gamma} + a'_{i\gamma} a'_{j\gamma}).$$

From (7.6.4) the derivatives of W are

$$\frac{\partial W}{\partial v_j} = 2n(t-1)v_j.$$

If we let

$$H_{ij} = \frac{1}{n(t-1)F} \sum_{\gamma} (a_{i\gamma} a_{j\gamma} + a'_{i\gamma} a'_{j\gamma}), \tag{7.6.8}$$

then (7.5.7) can be re-written as

$$\sum_i v_i H_{ij} = E^2 v_j. \tag{7.6.9}$$

Let us first verify that a solution of (7.6.9) will satisfy $\Sigma v_i = 0$. Summing both sides of (7.6.9) over j, and using (7.6.8), we get

$$\sum_i v_i = E^2 \sum_j v_j,$$

so that $\Sigma v_i = 0$ unless $E^2 = 1$, which can be ruled out in practice.

There is always a trivial solution of (7.6.9) for which E^2 is formally equal to unity. This is $v_j = 1$, all j, as is easily verified. To obtain the non-trivial solution let $\mathbf{v}$ be the column vector of the v_i, and let $\mathbf{H}$ be the $t \times t$ symmetric matrix of the H_{ij}. Then (7.6.9) becomes the matrix equation

$$\mathbf{H}\mathbf{v} = E^2\mathbf{v}, \tag{7.6.10}$$

showing that E^2 is an eigenvalue of $\mathbf{H}$, with $\mathbf{v}$ the corresponding eigenvector. Since we want the largest possible correlation ratio, we seek the largest of the non-trivial roots. As pointed out by Nishisato (1978), the trivial root is eliminated if $\mathbf{H}$ is replaced by $\mathbf{K} = \mathbf{H} - (1/t))\mathbf{J}$, where $\mathbf{J}$ is the $t \times t$ matrix of 1's. The numerical solution of $\mathbf{K}\mathbf{v} = E^2\mathbf{v}$ can be carried out by a simple iterative technique for obtaining the largest eigenvalue and the corresponding eigenvector.(*) Clearly, the latter is determined only up to a constant multiplier, so that it is convenient to impose the side condition $\mathbf{v}'\mathbf{v} = t$.

Guttman points out that his approach leads to scores which for the Thurstone–Mosteller model differ only by a linear transformation from the usual ones as obtained in §4.1. However, the optimal scaling method remains applicable when the assumptions of that model do not hold.

Nishisato (1978) has also proposed an alternative approach leading to the same results. In Nishisato (1980) this is further discussed in the setting of a general approach to optimal (or "dual") scaling for categorical data. Extensions to the treatment of ties, order effects, and missing values are indicated, and a review of previous contributions is provided.

7.7 Paradoxes

We conclude with a brief account of several results which at first sight seem contrary to intuition. As a simple example of this type, consider three players with preference probabilities

$$\pi_{12} = \pi > \tfrac{1}{2}, \quad \pi_{23} \geqslant \tfrac{1}{2}, \quad \pi_{13} = \max(\pi, \pi_{23}),$$

(*) See footnote on p. 105.

where we regard π as fixed. Ties are not allowed. Now suppose that, in this strong stochastic transitive ordering (A_1, A_2, A_3), player A_1 meets the winner of A_2 versus A_3. Then A_1 wins with probability

$$P_1 = \pi_{23}\pi + \pi_{32}\pi_{13} = \pi + \pi_{32}(\pi_{13} - \pi).$$

Thus $P_1 \geqslant \pi$, equality occurring only if $\pi_{32} = 0$ or if $\pi_{13} = \pi$, i.e., if $\pi_{32} = 1 - \pi$. In the range $0 \leqslant \pi_{32} \leqslant 1 - \pi$ we have

$$P_1 = \pi + \pi_{32}(1 - \pi_{32} - \pi),$$

which *increases* as π_{32} increases from 0 to $\frac{1}{2}(1-\pi)$. So as A_3 increases in strength, in the stated range, A_1's chances of winning increase; or stated more paradoxically, stronger opposition makes winning easier for A_1. The explanation lies, of course, in A_1's increasing probability of meeting A_3 rather than the substantially stronger A_2. Exercise 7.4 provides a related paradox.

The foregoing example may be regarded as a mini-tournament in which A_1 is a seeded player. In an interesting paper Hwang (1982) has re-examined seeding in Knock-out tournaments. It is easy to see that if the preference matrix (π_{ij}) is not restricted, then the player seeded highest may not have the largest probability of winning the tournament. Hwang shows that this paradoxical situation may continue to hold even when the π_{ij} are subject to strong stochastic transitivity and *all* players are seeded.

The standard method of running a KO tournament with seeds may be defined recursively for $t = 2^p$ players, labelled $1, 2, \ldots, t$ according to their seeding, 1 being best here. For $p = 1$, the pairing is 1, 2. Suppose $k_1, k_2; \ldots; k_{\frac{1}{2}t-1}, k_{\frac{1}{2}t}$ gives the schedule for 2^{p-1} players. Then the schedule for 2^p players is $k_1, t+1-k; \ldots; k_{\frac{1}{2}t}, t+1-k_{\frac{1}{2}t}$. Thus for $t = 4, 8$ the schedules are 1, 4; 2, 3 and 1, 8; 4, 5; 2, 7; 3, 6. If t is not a power of 2, dummy players with no chance of beating real players can be inserted and given the missing higher ranks to create a power of 2.

Consider the following preference matrix:

It is easily verified that, as $\varepsilon \to 0$, players 1 and 2 have respective probabilities $\frac{1}{4}$ and $\frac{3}{8}$ of winning the tournament. The anomaly is due to the fact that player 2 has probability $\frac{1}{2}$ of meeting 6 in the second round, a certain win.

The remedy proposed by Hwang is simple: After each round of the tournament rank the players according to their labels and use

Table 7.2 A preference matrix with strong stochastic transitivity

i \ j	2	3	4	5	6	7	8	
1	π	π	π	π	1	1	1	
2		π	π	π	1	1	1	
3			π	π	π	1	1	$\pi = 0{\cdot}5 + \varepsilon$
4				π	π	1	1	
5					π	1	1	
6						1	1	
7							1	

the traditional method of scheduling on the relabelled players. Thus in each round the best surviving player will be paired against the worst surviving player. This will clearly maximize the top seed's chances of winning the tournament (if one can rely on the initial seedings and their constancy).

Next, consider two Knock-out tournament plans that are identical except for an interchange of the starting positions of A_i and A_j. Chung and Hwang (1978) show that under the Bradley–Terry model the stronger player A_i $(\pi_i > \pi_j)$ has a better chance of winning the tournament. On averaging over the $t!$ assignments of players to starting positions, it is seen that A_i continues to have the better chance when the assignments are made at random. These results are not at all surprising, but the ingenious proof leaves open the question whether similar results hold for other models. Thus one might think that strong stochastic transitivity of the preference matrix should suffice: If A_i is stronger than A_j, then A_i should have the better chance of winning a tournament with random assignments. However, a counter-example has been given by Israel (1981). See also Hwang and Hsuan (1980).

EXERCISES

7.1 Show that a correct response with the duo–trio test is obtained if

either $\quad y_2 - y_1' < 0 \quad$ and $\quad y_1' + y_2 - 2y_1 < 0,$

or $\qquad y_2 - y_1' > 0 \quad \text{and} \quad y_1' + y_2 - 2y_1 > 0.$

Hence establish (7.1.3).

7.2 If y_1, y_1', y_2 are independent rectangular variates, the first two in the range $(\mu_1 - \sqrt{3}\sigma, \mu_1 + \sqrt{3}\sigma)$ and the last in $(\mu_2 - \sqrt{3}\sigma, \mu_2 + \sqrt{3}\sigma)$, show that

$$P_P = \tfrac{1}{2} + \frac{\delta}{2\sqrt{3}} - \frac{\delta^2}{24} \qquad 0 \leqslant \delta < 2\sqrt{3},$$

$$= 1 \qquad 2\sqrt{3} \leqslant \delta;$$

$$P_\Delta = \tfrac{1}{3} + \frac{\delta^2}{12} - \frac{\delta^3}{48\sqrt{3}} \qquad 0 \leqslant \delta < 2\sqrt{3},$$

$$= -\tfrac{1}{3} + \frac{\delta}{\sqrt{3}} - \frac{\delta^2}{12} + \frac{\delta^3}{144\sqrt{3}} \qquad 2\sqrt{3} \leqslant \delta < 4\sqrt{3},$$

$$= 1 \qquad 4\sqrt{3} \leqslant \delta;$$

$$P_D = \tfrac{1}{2} + \frac{\delta^2}{12} - \frac{7\delta^3}{288\sqrt{3}} \qquad 0 \leqslant \delta < 2\sqrt{3},$$

$$= \tfrac{1}{3} + \frac{\delta}{2\sqrt{3}} - \frac{\delta}{24} + \frac{\delta^3}{288\sqrt{3}} \qquad 2\sqrt{3} \leqslant \delta < 4\sqrt{3},$$

$$= 1 \qquad 4\sqrt{3} \leqslant \delta;$$

where $\delta = (\mu_2 - \mu_1)/\sigma$.

(Ura, 1960a)

7.3 In an experiment on $A_1, A_2, \ldots, A_t$, with n_{ij} comparisons of A_i and A_j $(1 \leqslant i < j \leqslant t)$, let w_i, t_i be the total number of wins, ties for A_i. If $a_i = w_i + \frac{1}{2}t_i$ and $T = \frac{1}{2}\Sigma t_i$, show that the ML estimates $(\mathbf{p}, \hat{\nu})$ of the parameter vector $(\boldsymbol{\pi}, \nu)$ in Davidson's model (7.3.4–5) are solutions of the equations

$$a_i/p_i - g_i(\mathbf{p}, \hat{\nu}) = 0 \quad i = 1, 2, \ldots, t,$$

$$T/\hat{\nu} - h(\mathbf{p}, \hat{\nu}) = 0,$$

where

$$g_i(\mathbf{p}, \hat{\nu}) = \sum_j n_{ij}(1 + \tfrac{1}{2}\hat{\nu}\sqrt{p_j/p_i}) \Big/ (p_i + p_j + \hat{\nu}\sqrt{p_i p_j}),$$

$$h(\mathbf{p}, \hat{\nu}) = \sum_{i<j}\sum n_{ij}\sqrt{p_i p_j}\Big/(p_i + p_j + \hat{\nu}\sqrt{p_i p_j}).$$

Hence show that for $n_{ij} = n$ (all $i \neq j$) the p_i rank the A_i according to the scores a_i.

(Davidson, 1970)

7.4 Suppose the merits of the players A_1, A_2, A_3 satisfy the Bradley–Terry model (1.3.10) with $\pi_1 \geqslant \pi_2 \geqslant \pi_3$. Show that A_1 beats the winner of A_2 versus A_3 with probability

$$P_1 = \frac{\pi_1}{1-\pi_1}\left(\frac{\pi_3}{\pi_1+\pi_3} + \frac{1-\pi_1-\pi_3}{1-\pi_3}\right),$$

and hence that, for π_1 fixed, P_1 increases with π_3 as long as $\pi_3 < \pi_2$.

APPENDIX

Appendix Table 1 Critical values for the sum of squares of scores

$$\sum_{i=1}^{t} a_i^2$$

Based on Trawinski (1961) by permission of the author

$t=3$		$t=3$		$t=4$		$t=5$	
$n=3$	45 —	$n=12$	488 518	$n=2$	52 56	$n=2$	104 110
4	72 80	13	569 603	3	106 114	3	216 228
5	101 135	14	660 692	4	176 188	$n=1$	
6	140 150	15	747 779	5	266 282	$t=6$	55 —
7	185 197	16	846 882	6	372 392	7	85 89
8	234 248	17	945 989	7	498 522	8	126 132
9	285 305	18	1058 1098			9	178 186
10	350 372	19	1179 1211			10	245 253
11	417 441	20	1296 1346				

Entries for $n=1$, $t=9$, 10 are derived from Alway (1962b). For $n=1$, $t \leqslant 10$ a table of the cdf of Σa_i^2 is given by Gridgeman (1963).

Appendix Table 2 Critical values for the difference between scores of two pre-assigned objects

From Starks and David (1961) by permission of the authors and the Editor of *Biometrika*

Experiment size		$\alpha = 0{\cdot}01$		$\alpha = 0{\cdot}05$	
n	t	One-sided test m'_c	Two-sided test m_c	One-sided test m'_c	Two-sided test m_c
1	$\leqslant 4$	No significant values		No significant values	
1	5	4	None possible	4	4
1	6	5	5	4	4
1	7	5	5	4	5
1	8	5	6	4	5
1	9	6	6	4	5
1	10	6	7	5	5
1	11	6	7	5	6
1	12	7	7	5	6
1	13	7	7	5	6
1	14	7	8	5	6
1	15	7	8	5	6
1	16	7	8	6	6
2	3	No significant values		4	4
2	4	5	6	4	5
2	5	6	6	5	5
3	3	6	6	4	5
3	4	6(*)	7	5	6
4	3	6(†)	7	5	6
4	4	7	8	6	6
All larger values of n or t		m'_c = smallest integer $\geqslant 2{\cdot}33\sigma + 0{\cdot}5$	m_c = smallest integer $\geqslant 2{\cdot}56\sigma + 0{\cdot}5$	m'_c = smallest integer $\geqslant 1{\cdot}64\sigma + 0{\cdot}5$	m_c = smallest integer $\geqslant 1{\cdot}96\sigma + 0{\cdot}5$

$[\sigma = (\tfrac{1}{2}nt)^{1/2}]$

(*) $P_{436} = 0.0103$.
(†) $P_{346} = 0{\cdot}010\,01$.

Appendix Table 3 Critical values $R_{\beta(\alpha)}$ for the multiple comparison range test(*)

From Starks and David (1961) by permission of the authors and the Editor of *Biometrika*

		$\alpha = 0{\cdot}01$		$\alpha = 0{\cdot}05$	
t	n	$R_{\beta(\alpha)}$	β	$R_{\beta(\alpha)}$	β
3	1	None	None	None	None
3	2	None	None	None	None
3	3	6	0·01	None	None
3	4	7	·01	6	0·05
3	5	8	·01	7	·04
3	6	9	0·01	8	0·03
3	7	10	·01	8	·05
3	8	10	·01	9	·03
3	9	11	·01	9	·05
3	10	12	·01	10	·03
4	1	None	None	None	None
4	2	6	0·01	None	None
4	3	8	·005	7	0·03
4	4	9	·01	8	·03
4	5	10	·01	9	·03
4	6	11	0·01	9	0·06
4	7	12	·01	10	·05
4	8	13	·01	11	·04
5	1	None	None	None	None
5	2	7	·0154	6	·08
5	3	9	0·01	8	0·04
5	4	11	·005	9	·05
5	5	12	·01	10	·05
6	1	None	None	5	·0586
7	1	None	None	6	·0205
8	1	7	·0068	6	·0738

(*) These critical values $R_{\beta(\alpha)}$ have been chosen in such a way as to make β as close to α as possible, rather than requiring that $R_{\beta(\alpha)}$ be such that $\beta \leqslant \alpha$, as was done in the defining equation (3.7.1).

Appendix Table 4 Smallest number of replications required to ensure with at least a predetermined probability P' the selection of the best object when

$$\Pr(A_{(t)} \to A_{(i)}) \geqslant \pi (i = 1, 2, \ldots, t-1), \Pr(A_{(i)} \to A_{(j)}) = \tfrac{1}{2}$$
$$(i \neq j; i, j = 1, 2, \ldots, t-1)$$

(a) $P' = 0{\cdot}75$

t \ π	0·55	0·60	0·65	0·70	0·75	0·80	0·85	0·90	0·95
2	45	11	5	3	1	1	1	1	1
3	69	17	8	4	3	2	1	1	1
4	71	18	8	5	3	2	2	1	1
5	68	17	8	4	3	2	2	1	1
6	65	16	7	4	3	2	2	1	1
7	61	15	7	4	3	2	2	1	1
8	58	15	7	4	3	2	1	1	1
9	54	14	6	4	2	2	1	1	1
10	52	13	6	4	2	2	1	1	1
12	47	12	5	3	2	2	1	1	1
14	43	11	5	3	2	2	1	1	1
16	39	10	5	3	2	1	1	1	1
18	37	9	4	3	2	1	1	1	1
20	34	9	4	3	2	1	1	1	1

(b) $P' = 0{\cdot}90$

t \ π	0·55	0·60	0·65	0·70	0·75	0·80	0·85	0·90	0·95
2	163	41	17	9	7	5	3	1	1
3	165	41	18	10	6	4	3	2	1
4	150	37	16	9	6	4	3	2	1
5	135	33	15	8	5	4	3	2	1
6	122	30	13	7	5	3	2	2	1
7	112	28	12	7	4	3	2	2	1
8	103	26	11	6	4	3	2	2	1
9	95	24	11	6	4	3	2	1	1
10	89	22	10	6	4	3	2	1	1
12	79	20	9	5	3	2	2	1	1
14	71	18	8	5	3	2	2	1	1
16	64	16	7	4	3	2	2	1	1
18	59	15	7	4	3	2	1	1	1
20	55	14	6	4	2	2	1	1	1

Appendix Table 4 (*concluded*)

(c) $P' = 0{\cdot}95$

t \ π	0·55	0·60	0·65	0·70	0·75	0·80	0·85	0·90	0·95
2	269	67	29	15	9	7	5	3	1
3	243	60	26	15	9	6	4	3	2
4	212	52	23	12	8	5	4	3	2
5	186	46	20	11	7	5	3	3	2
6	166	41	18	10	6	4	3	2	2
7	150	37	16	9	6	4	3	2	2
8	137	34	15	8	5	4	3	2	2
9	126	31	14	8	5	3	2	2	1
10	117	29	13	7	5	3	2	2	1
12	102	26	11	6	4	3	2	2	1
14	91	23	10	6	4	3	2	1	1
16	82	21	9	5	3	2	2	1	1
18	75	19	9	5	3	2	2	1	1
20	69	17	8	5	3	2	2	1	1

(d) $P' = 0{\cdot}99$

t \ π	0·55	0·60	0·65	0·70	0·75	0·80	0·85	0·90	0·95
2	537	133	57	31	19	13	9	5	3
3	433	106	45	24	15	10	7	5	3
4	358	88	38	20	12	8	6	4	3
5	306	75	33	18	11	7	5	4	3
6	267	66	29	16	10	6	4	3	2
7	238	59	26	14	9	6	4	3	2
8	214	53	23	13	8	5	4	3	2
9	195	48	21	12	7	5	3	2	2
10	180	45	20	11	7	4	3	2	2
12	155	39	17	9	6	4	3	2	2
14	137	34	15	8	5	4	3	2	1
16	123	31	14	8	5	3	2	2	1
18	112	28	12	7	4	3	2	2	1
20	102	26	11	6	4	3	2	2	1

Appendix Table 5 Values of ν for the decision rule $\mathscr{R}$ of §6.3

(a) $P^* = 0{\cdot}75$

n \ t	2	3	4	5	6	7	8	9	10
1	1	1	2	2	2	3	3	3	4
2	0	2	2	3	3	4	4	5	5
3	1	2	3	4	4	5	5	6	6
4	2	2	3	4	5	5	6	7	7
5	1	3	4	5	5	6	7	7	8
6	2	3	4	5	6	7	7	8	9
7	1	3	4	5	6	7	8	9	9
8	2	4	5	6	7	8	9	9	10
9	3	4	5	6	7	8	9	10	11
10	2	4	5	7	8	9	10	10	11
11	3	4	6	7	8	9	10	11	12
12	2	4	6	7	8	9	10	11	12
13	3	5	6	7	9	10	11	12	13
14	2	5	6	8	9	10	11	12	13
15	3	5	7	8	9	11	12	13	14
16	2	5	7	8	10	11	12	13	14
17	3	5	7	9	10	11	12	14	15
18	2	5	7	9	10	12	13	14	15
19	3	5	7	9	10	12	13	14	16
20	4	6	8	9	11	12	14	15	16
25	3	6	8	10	12	14	15	17	18
30	4	7	9	11	13	15	17	18	20
35	3	7	10	12	14	16	18	20	21
40	4	8	11	13	15	17	19	21	23
45	5	8	11	14	16	18	20	22	24
50	4	9	12	15	17	19	21	23	25
60	6	10	13	16	19	21	23	26	28
70	6	10	14	17	20	23	25	28	30
80	6	11	15	18	22	24	27	30	32
90	6	12	16	20	23	26	29	31	34
100	6	12	17	21	24	27	30	33	36

Appendix Table 5 (*continued*)

(a) $P^* = 0{\cdot}75$

n \ t	11	12	13	14	15	16	17	18	19	20
1	4	4	4	5	5	5	5	5	6	6
2	5	6	6	6	7	7	7	8	8	8
3	7	7	7	8	8	9	9	9	10	10
4	8	8	9	9	10	10	10	11	11	12
5	9	9	10	10	11	11	12	12	13	13
6	9	10	11	11	12	12	13	13	14	14
7	10	11	11	12	13	13	14	14	15	15
8	11	12	12	13	14	14	15	15	16	16
9	12	12	13	14	14	15	16	16	17	17
10	12	13	14	14	15	16	16	17	18	18
11	13	14	14	15	16	17	17	18	19	19
12	13	14	15	16	17	17	18	19	19	20
13	14	15	16	16	17	18	19	20	20	21
14	14	15	16	17	18	19	19	20	21	22
15	15	16	17	18	18	19	20	21	22	23
16	15	16	17	18	19	20	21	22	22	23
17	16	17	18	19	20	21	21	22	23	24
18	16	17	18	19	20	21	22	23	24	25
19	17	18	19	20	21	22	23	24	24	25
20	17	18	19	20	21	22	23	24	25	26
25	19	20	22	23	24	25	26	27	28	29
30	21	22	24	25	26	27	29	30	31	32
35	23	24	26	27	28	30	31	32	33	34
40	24	26	27	29	30	32	33	34	36	37
45	26	27	29	31	32	34	35	36	38	39
50	27	29	31	32	34	35	37	38	40	41
60	30	32	33	35	37	39	40	42	44	45
70	32	34	36	38	40	42	44	45	47	49
80	34	37	39	41	43	45	47	48	50	52
90	36	39	41	43	45	47	49	51	53	55
100	38	41	43	46	48	50	52	54	56	58

Appendix Table 5 (*continued*)

(b) $P^* = 0{\cdot}90$

n \ t	2	3	4	5	6	7	8	9	10
1	1	2	2	3	3	4	4	4	5
2	2	3	3	4	5	5	6	6	7
3	3	3	4	5	6	6	7	8	8
4	2	4	5	6	7	7	8	9	9
5	3	4	5	6	7	8	9	10	11
6	4	5	6	7	8	9	10	11	12
7	3	5	6	8	9	10	11	12	12
8	4	5	7	8	9	10	11	12	13
9	3	6	7	9	10	11	12	13	14
10	4	6	8	9	10	12	13	14	15
11	5	6	8	10	11	12	13	15	16
12	4	7	8	10	11	13	14	15	16
13	5	7	9	10	12	13	15	16	17
14	4	7	9	11	12	14	15	16	18
15	5	8	9	11	13	14	16	17	18
16	6	8	10	12	13	15	16	18	19
17	5	8	10	12	14	15	17	18	19
18	6	8	10	12	14	16	17	19	20
19	5	8	11	13	14	16	18	19	21
20	6	9	11	13	15	17	18	20	21
25	7	10	12	15	17	18	20	22	24
30	8	11	13	16	18	20	22	24	26
35	7	11	15	17	20	22	24	26	28
40	8	12	16	18	21	23	26	28	30
45	9	13	16	19	22	25	27	29	32
50	10	14	17	21	23	26	29	31	33
60	10	15	19	23	26	29	31	34	37
70	12	16	21	24	28	31	34	37	39
80	12	17	22	26	30	33	36	39	42
90	12	18	23	28	31	35	38	42	45
100	12	19	25	29	33	37	41	44	47

Appendix Table 5 (*continued*)

(b) $P^* = 0{\cdot}90$

n \ t	11	12	13	14	15	16	17	18	19	20
1	5	5	6	6	6	6	7	7	7	7
2	7	8	8	8	9	9	9	10	10	10
3	9	9	10	10	11	11	12	12	12	13
4	10	11	11	12	12	13	13	14	14	15
5	11	12	12	13	14	14	15	15	16	16
6	12	13	14	14	15	16	16	17	17	18
7	13	14	15	16	16	17	18	18	19	20
8	14	15	16	17	17	18	19	20	20	21
9	15	16	17	18	18	19	20	21	21	22
10	16	17	18	19	19	20	21	22	23	23
11	17	18	18	19	20	21	22	23	24	24
12	17	18	19	20	21	22	23	24	25	26
13	18	19	20	21	22	23	24	25	26	27
14	19	20	21	22	23	24	25	26	27	28
15	19	21	22	23	24	25	26	27	28	29
16	20	21	22	24	25	26	27	28	29	29
17	21	22	23	24	25	26	27	28	29	30
18	21	23	24	25	26	27	28	29	30	31
19	22	23	24	26	27	28	29	30	31	32
20	22	24	25	26	27	29	30	31	32	33
25	25	27	28	29	31	32	33	34	36	37
30	28	29	30	32	34	35	36	38	39	40
35	30	31	33	35	36	38	39	41	42	44
40	32	34	35	37	39	40	42	44	45	47
45	34	36	37	39	41	43	45	46	48	49
50	36	38	39	42	43	45	47	49	50	52
60	39	41	43	46	48	50	52	53	55	57
70	42	44	46	49	51	54	56	58	60	62
80	45	48	50	53	55	57	60	62	64	66
90	48	50	53	56	58	61	63	65	68	70
100	50	53	56	59	61	64	67	69	71	74

Appendix Table 5 (*continued*)

(c) $P^* = 0{\cdot}95$

n \ t	2	3	4	5	6	7	8	9	10
1	1	2	3	3	4	4	5	5	5
2	2	3	4	5	5	6	7	7	8
3	3	4	5	6	7	7	8	9	9
4	4	5	6	7	8	9	9	10	11
5	3	5	6	8	9	10	10	11	12
6	4	6	7	8	9	10	11	12	13
7	5	6	8	9	10	11	12	13	14
8	4	7	8	10	11	12	13	14	15
9	5	7	9	10	12	13	14	15	16
10	6	7	9	11	12	14	15	16	17
11	5	8	10	11	13	14	16	17	18
12	6	8	10	12	13	15	16	17	19
13	5	8	11	12	14	15	17	18	19
14	6	9	11	13	14	16	18	19	20
15	7	9	11	13	15	17	18	20	21
16	6	9	12	14	15	17	19	20	22
17	7	10	12	14	16	18	19	21	22
18	6	10	12	14	16	18	20	21	23
19	7	10	13	15	17	19	20	22	24
20	8	10	13	15	17	19	21	23	24
25	9	12	15	17	19	21	23	25	27
30	8	13	16	19	21	23	26	28	30
35	9	14	17	20	23	25	28	30	32
40	10	15	18	22	24	27	30	32	34
45	11	16	20	23	26	29	31	34	36
50	12	17	21	24	27	30	33	36	38
60	12	18	23	26	30	33	36	39	42
70	14	20	24	29	32	36	39	42	45
80	14	21	26	31	35	38	42	45	48
90	16	22	28	32	37	41	44	48	51
100	16	23	29	34	39	43	47	51	54

Appendix Table 5 (*continued*)

(c) $P^* = 0{\cdot}95$

n \ t	11	12	13	14	15	16	17	18	19	20
1	6	6	6	7	7	7	8	8	8	8
2	8	9	9	9	10	10	11	11	11	12
3	10	10	11	12	12	13	13	14	14	14
4	11	12	13	13	14	15	15	16	16	17
5	13	14	14	15	16	16	17	17	18	19
6	14	15	16	16	17	18	18	19	20	20
7	15	16	17	18	18	19	20	21	21	22
8	16	17	18	19	20	21	21	22	23	24
9	17	18	19	20	21	22	23	23	24	25
10	18	19	20	21	22	23	24	25	25	26
11	19	20	21	22	23	24	25	26	27	28
12	20	21	22	23	24	25	26	27	28	29
13	21	22	23	24	25	26	27	28	29	30
14	21	23	24	25	26	27	28	29	30	31
15	22	23	25	26	27	28	29	30	31	32
16	23	24	26	27	28	29	30	31	32	33
17	24	25	26	28	29	30	31	32	33	34
18	24	26	27	28	30	31	32	33	34	35
19	25	26	28	29	30	32	33	34	35	36
20	26	27	29	30	31	32	34	35	36	37
25	29	30	32	33	35	36	38	39	40	42
30	31	33	35	37	38	40	41	43	44	46
35	34	36	38	40	41	43	45	46	48	49
40	36	38	40	42	44	46	48	49	51	53
45	39	41	43	45	47	49	51	52	54	56
50	41	43	45	47	49	51	53	55	57	59
60	44	47	49	52	54	56	58	60	62	64
70	48	51	53	56	58	61	63	65	67	70
80	51	54	57	60	62	65	67	70	72	74
90	54	58	60	63	66	69	71	74	76	79
100	57	61	64	67	70	73	75	78	81	83

Appendix Table 5 (*continued*)

(d) $P^* = 0{\cdot}99$

n \ t	2	3	4	5	6	7	8	9	10
1	1	2	3	4	4	5	6	6	7
2	2	4	5	6	7	8	8	9	9
3	3	5	6	8	9	9	10	11	12
4	4	6	7	9	10	11	12	13	13
5	5	7	8	10	11	12	13	14	15
6	6	7	9	11	12	13	14	15	16
7	5	8	10	12	13	14	16	17	18
8	6	9	11	12	14	15	17	18	19
9	7	9	11	13	15	16	18	19	20
10	8	10	12	14	16	17	19	20	21
11	7	10	13	15	16	18	19	21	22
12	8	11	13	15	17	19	20	22	23
13	9	11	14	16	18	19	21	23	24
14	8	12	14	16	18	20	22	24	25
15	9	12	15	17	19	21	23	24	26
16	10	12	15	18	20	22	23	25	27
17	9	13	16	18	20	22	24	26	28
18	10	13	16	19	21	23	25	27	28
19	9	14	17	19	21	24	26	27	29
20	10	14	17	20	22	24	26	28	30
25	11	16	19	22	25	27	29	31	34
30	12	17	21	24	27	30	32	34	37
35	13	19	22	26	29	32	35	37	40
40	14	20	24	28	31	34	37	40	42
45	15	21	25	29	33	36	39	42	45
50	16	22	27	31	35	38	41	45	47
60	18	24	29	34	38	42	45	49	52
70	20	26	32	37	41	45	49	53	56
80	20	28	34	39	44	48	52	56	60
90	22	30	36	42	47	51	56	60	64
100	24	31	38	44	49	54	59	63	67

Appendix Table 5 (*concluded*)

(d) $P^* = 0{\cdot}99$

n \ t	11	12	13	14	15	16	17	18	19	20
1	7	7	8	8	9	9	9	10	10	10
2	10	11	11	12	12	13	13	13	14	14
3	12	13	14	14	15	15	16	16	17	18
4	14	15	16	16	17	18	18	19	20	20
5	16	17	18	18	19	20	21	21	22	23
6	17	18	19	20	21	22	23	23	24	25
7	19	20	21	22	23	23	24	25	26	27
8	20	21	22	23	24	25	26	27	28	29
9	21	22	24	25	26	27	28	29	29	30
10	22	24	25	26	27	28	29	30	31	32
11	24	25	26	27	28	29	30	32	33	34
12	25	26	27	28	30	31	32	33	34	35
13	26	27	28	30	31	32	33	34	35	36
14	27	28	29	31	32	33	34	36	37	38
15	28	29	30	32	33	34	36	37	38	39
16	28	30	31	33	34	35	37	38	39	40
17	29	31	32	34	35	37	38	39	40	42
18	30	32	33	35	36	38	39	40	42	43
19	31	33	34	36	37	39	40	41	43	44
20	32	33	35	37	38	40	41	43	44	45
25	36	37	39	41	43	44	46	48	49	51
30	39	41	43	45	47	49	50	52	54	55
35	42	44	46	49	51	52	54	56	58	60
40	45	47	50	52	54	56	58	60	62	64
45	48	50	53	55	57	60	62	64	66	68
50	50	53	55	58	60	63	65	67	69	72
60	55	58	61	64	66	69	71	74	76	78
70	59	63	66	69	71	74	77	80	82	85
80	64	67	70	73	76	79	82	85	88	90
90	67	71	74	78	81	84	87	90	93	96
100	71	75	78	82	85	89	92	95	98	101

REFERENCES

An extensive bibliography up to 1976 is given in Davidson and Farquhar (1976).

Abelson, R. M. and Bradley, R. A. (1954). A 2 × 2 factorial with paired comparisons. *Biometrics* **10**, 487–502.

Alway, G. G. (1962a). Matrices and sequences. *Math. Gaz.* **46**, 208–13.

Alway, G. G. (1962b). The distribution of the number of circular triads in paired comparisons. *Biometrika* **49**, 265–9.

Anderson, T. W. (1984). *Introduction to Multivariate Statistical Analysis*, 2nd edn. New York: Wiley.

Archbold, J. W. and Johnson, N. L. (1958). A construction for Room's squares and an application in experimental design. *Ann. Math. Statist.* **29**, 219–25.

Arya, A. S. (1983). Circulant plans for partial diallel crosses. *Biometrics* **39**, 43–52.

Atkinson, A. C. (1972). A test of the linear logistic and Bradley–Terry models. *Biometrika* **59**, 37–42.

Bauer, D. F. (1978). Circular triads when not all paired comparisons are made. *Biometrics* **34**, 458–61.

Beaver, R. J. (1974). Locally asymptotically most stringent tests for paired comparison experiments. *J. Amer. Statist. Assoc.* **69**, 423–7.

Beaver, R. J. (1977a). Weighted least-squares analysis of several univariate Bradley–Terry models. *J. Amer. Statist. Assoc.* **72**, 629–34.

Beaver, R. J. (1977b). Weighted least squares response surface fitting in factorial paired comparisons. *Comm. Statist. A — Theory Methods A* **6**, 1275–87.

Beaver, R. J. and Gokhale, D. V. (1975). A model to incorporate within-pair order effects in paired comparisons. *Comm. Statist.* **4**, 923–39.

Beaver, R. J. and Rao, P. V. (1972). The use of limit theorems in paired and triple comparison model building. *J. Math. Psychol.* **9**, 92–103.

Beaver, R. J. and Rao, P. V. (1973). On ties in triple comparisons. *Trab. Estadist.* **24**, 77–92.

Bechhofer, R. E. (1954). A single-sample multiple decision procedure for ranking means of normal populations with known variances. *Ann. Math. Statist.* **35**, 16–39.

Bechhofer, R. E. (1970). On ranking the players in a 3-player tournament. In *Nonparametric Techniques in Statistical Inference*, ed. M. L. Puri. Cambridge: University Press, 545–59.

Bechhofer, R. E. and Kulkarni, R. V. (1982). Closed adaptive sequential procedures for selecting the best of $k \geqslant 2$ Bernoulli populations. *Proc. 3rd Purdue Symp. on Statistical Decision Theory and Related Topics,* ed. S. S. Gupta and J. Berger. New York: Academic Press, 61–108.

Bechtel, G. G. (1967). The analysis of variance and pairwise scaling. *Psychometrika* **32**, 47–65.

Berge, C. (1962). *The Theory of Graphs and its Applications,* transl. A. Doig. London: Methuen; New York: Wiley.

Berkum, E. E. M. van (1985). Optimal paired comparison designs for factorial and quadratic models. *J. Statist. Plann. Inference* **15**, 265–78.

Bezembinder, T. G. G. (1981). Circularity and consistency in paired comparisons. *Brit. J. Math. Statist. Psychol.* **34**, 16–37.

Bliss, C. I. (1960). Some statistical aspects of preference and related tests. *Appl. Statist.* **9**, 8–19.

Bliss, C. I., Greenwood, M. L., and White, E. S. (1956). A rankit analysis of paired comparisons for measuring the effect of sprays on flavor. *Biometrics* **12**, 381–403.

Block, H. D. and Marschak, J. (1960). Random orderings and stochastic theories of responses. *Contributions to Probability and Statistics,* ed. I. Olkin *et al.* Stanford: University Press, 97–132.

Bock, R. D. (1956). The selection of judges for preference testing. *Psychometrika* **21**, 349–66.

Bock, R. D. (1958). Remarks on the test of significance for the method of paired comparisons. *Psychometrika* **23**, 323–34.

Bock, R. D. and Jones, L. V. (1968). *The Measurement and Prediction of Judgment and Choice.* San Francisco: Holden-Day.

Bose, R. C. (1942). A note on the resolvability of balanced incomplete block designs. *Sankhyā* **6**, 105–10.

Bose, R. C. (1956). Paired comparison designs for testing concordance between judges. *Biometrika* **43**, 113–21.

Bose, R. C. and Cameron, J. M. (1965). The bridge tournament problem and calibration designs for comparing pairs of objects. *J. Res. Nat. Bur. Stand. B* **69**, 323–32.

Bradley, R. A. (1953). Some statistical methods in taste testing and quality evaluation. *Biometrics* **9**, 22–38.

Bradley, R. A. (1954a). Incomplete block rank analysis: on the appropriateness of the model for a method of paired comparisons. *Biometrics* **10**, 375–90.

Bradley, R. A. (1954b). The rank analysis of incomplete block designs. II. Additional tables for the method of paired comparisons. *Biometrika* **41**, 502–37.

Bradley, R. A. (1955). Rank analysis of incomplete block designs. III. Some large-sample results on estimation and power for a method of paired comparisons. *Biometrika* **42**, 450–70.

Bradley, R. A. (1963). Some relationships among sensory difference tests. *Biometrics* **19**, 385–97.

Bradley, R. A. (1976). Science, statistics and paired comparisons. *Biometrics* **32**, 213–32.

Bradley, R. A. (1984). Paired comparisons: some basic procedures and examples. *Handbook of Statistics* Vol. 4, ed. P. R. Krishnaiah and P. K. Sen. Amsterdam: North-Holland, 299–326.

Bradley, R. A. and El-Helbawy, A. T. (1976). Treatment contrasts in paired comparisons: basic procedures with application to factorials, *Biometrika* **63**, 255–62.

Bradley, R. A. and Harmon, T. J. (1964). The modified triangle test. *Biometrics* **20**, 608–25.

Bradley, R. A. and Terry, M. E. (1952). The rank analysis of incomplete block designs. I. The method of paired comparisons. *Biometrika* **39**, 324–45.

Brunk, H. D. (1960). Mathematical models for ranking from paired comparisons. *J. Amer. Statist. Assoc.* **55**, 503–20.

Bühlmann, H. and Huber, P. J. (1963). Pairwise comparison and ranking in tournaments. *Ann. Math. Statist.* **34**, 501–10.

Byer, A. J. and Abrams, D. (1953). A comparison of the triangular and two-sample taste test methods. *Food Technology* **7**, 185–7.

Carroll, Lewis (1947). Lawn tennis tournaments. *The Complete Works of Lewis Carroll.* New York: Modern Library.

Chakravarti, I. M. (1976). Statistical designs from Room's squares with applications, *Experientia Supplementum* **22**, *Contributions to Applied Statistics.* Ed. W. J. Ziegler. Basel: Birkhäuser, 223–31.

Chang-Li-Chien (1961). The maximum and minimum probabilities of cyclic stochastic inequalities. *Acta Math. Sinica* **11**, No. 3, transl. in *Chinese Mathematics* **2**, 279–83.

Chen, C. and Smith, T. M. (1984). A Bayes-type estimator for the Bradley–Terry model for paired comparison. *J. Statist. Plann. Inference* **10**, 9–14.

Chung, F. R. D. and Hwang, F. K. (1978). Do stronger players win more knock-out tournaments? *J. Amer. Statist. Assoc.* **73**, 593–6.

Clatworthy, W. H. (1955). Partially balanced incomplete block designs with two associate classes and two treatments per block. *J. Res. Nat. Bur. Stand.* **54**, 177–90.

Clatworthy, W. H. (1973). Tables of two-associate-class partially balanced designs. *Nat. Bur. Stand. Appl. Math. Ser.* 63 (Washington, D.C.).

Cochran, W. G. and Cox, G. M. (1957). *Experimental Designs,* 2nd edn. New York: Wiley.

Coombs, C. H. (1959). Inconsistency of preferences as a measure of psychological distance. *Measurement, Definitions and Theories,* ed. C. W. Churchman and P. Ratoosh. New York: Wiley, 221–32.

Coombs, C. H. (1964). *A Theory of Data.* New York: Wiley.

Cowden, D. J. (1975). A method of evaluating contestants. *Amer. Statist.* **29**, 82–4.

Cramér, H. (1946). *Mathematical Methods of Statistics.* Princeton: University Press.

Daley, D. J. (1979). Markov chains and a pecking order problem. *Interactive Statistics,* ed. D. McNeil. New York: Wiley, 247–54.

Daniels, H. E. (1969). Round-Robin tournament scores. *Biometrika* **56**,

295–9.

David, F. N. and Barton, D. E. (1962). *Combinatorial Chance.* London: Griffin; New York: Hafner.

David, H. A. (1959). Tournaments and paired comparisons. *Biometrika* **46**, 139–49.

David, H. A. (1963). The structure of cyclic paired-comparison designs. *J. Aust. Math. Soc.* **3**, 117–27.

David, H. A. (1967). Resolvable cyclic designs. *Sankhyā* **29**, 191–8.

David, H. A. (1971). Ranking the players in a Round Robin tournament. *Rev. Int. Statist. Inst.* **39**, 137–47.

David, H. A. (1972). Enumeration of cyclic graphs and cyclic designs. *J. Comb. Theory* **13**, 303–8.

David, H. A. (1987). Ranking from unbalanced paired-comparison data. *Biometrika* **74**, 432–6.

David, H. A. and Andrews, D. M. (1987). Closed adaptive sequential paired-comparison selection procedures. *J. Statist. Comput. Simul.* **27**, 127–41.

David, H. A. and Trivedi, M. C. (1962). Pair, triangle, and duo-trio tests. Virginia Poly. Inst., Tech. Rep. 54.

Davidson, R. R. (1969). On a relationship between two representations of a model for paired comparisons. *Biometrics* **25**, 597–600.

Davidson, R. R. (1970). On extending the Bradley–Terry model to accommodate ties in paired comparison experiments. *J. Amer. Statist. Assoc.* **65**, 317–28.

Davidson, R. R. (1973). Ranking by maximum likelihood under a model for paired comparisons. *Comm. Statist.* **1**, 381–91.

Davidson, R. R. and Beaver, R. J. (1977). On extending the Bradley–Terry model to incorporate within-pair order effects. *Biometrics* **33**, 693–702.

Davidson, R. R. and Bradley, R. A. (1969). Multivariate paired comparisons: the extension of a univariate model and associated estimation and test procedures. *Biometrika* **56**, 81–95.

Davidson, R. R. and Bradley, R. A. (1970). Multivariate paired comparisons: some large-sample results on estimation and tests of equality of preference. In *Nonparametric Techniques in Statistical Inference*, ed. M. L. Puri. Cambridge: University Press, 111–25.

Davidson, R. R. and Farquhar, P. H. (1976). A bibliography on the method of paired comparisons. *Biometrics* **32**, 241–52.

Davidson, R. R. and Solomon, D. L. (1973). A Bayesian approach to paired comparison experimentation. *Biometrika* **60**, 477–87.

Davis, R. L. (1954). Structures of dominance relations. *Bull. Math. Biophysics* **16**, 131–40.

Davoodzadeh, J. and Beaver, R. J. (1982). Models for multivariate paired comparison experiments with ties. *J. Math. Psychol.* **25**, 269–81.

Debreu, G. (1960). Review of R. D. Luce, Individual choice behavior: a theoretical analysis. *Amer. Econ. Rev.* **50**, 186–8.

DeCani, J. S. (1969). Maximum likelihood paired comparison ranking by linear programming. *Biometrika* **56**, 537–45.

DeCani, J. S. (1972). A branch and bound algorithm for maximum likelihood paired comparison ranking. *Biometrika* **59**, 131–6.

DeSoete, G., Carroll, J. D., and DeSarbo, W. S. (1986). The wandering ideal point model: a probabilistic multidimensional unfolding model for paired comparisons data. *J. Math. Psychol.* **30**, 28–41.

Draper, N. R., Hunter, W. G., and Tierney, D. E. (1969). Which product is better? *Technometrics* **11**, 309–20.

Duncan, D. B. (1955). Multiple range and multiple *F* tests. *Biometrics* **11**,1–42.

Durbin, J. (1951). Incomplete blocks in ranking experiments. *Brit. J. Psychol.* (Statist. Sect.) **4**, 85–90.

Dykstra, O. (1956). A note on the rank analysis of incomplete block designs – applications beyond the scope of existing tables. *Biometrics* **12**, 301–6.

Dykstra, O. (1958). Factorial experimentation in Scheffé's analysis of variance for paired comparisons. *J. Amer. Statist. Assoc.* **53**, 529–42.

Dykstra, O. (1960). Rank analysis of incomplete block designs: a method of paired comparisons employing unequal repetitions on pairs. *Biometrics* **16**, 176–88.

Ehrenberg, A. S. C. (1952). On sampling from a population of rankers. *Biometrika* **39**, 82–7.

El-Helbawy, A. T. (1984) Asymptotic relative efficiency of designs for factorial paired comparison experiments. *J. Statist. Plann. Inference* **10**, 105–13.

El-Helbawy, A. T. and Ahmed, E. A. (1984). Optimal design results for 2^n factorial paired comparison experiments. *Comm. Statist. A – Theory Methods* **13**, 2827–45.

El-Helbawy, A. T. and Bradley, R. A. (1978). Treatment contrasts in paired comparisons: large-sample results, applications, and some optimal designs. *J. Amer. Statist. Assoc.* **73**, 831–9.

Ennis, D. M. and Mullen, K. (1986a). A multivariate model for discrimination methods. *J. Math. Psychol.* **30**, 206–19.

Ennis, D. M. and Mullen, K. (1986b). Theoretical aspects of sensory discrimination. *Chemical Senses* **11**, 513–22.

Falmagne, J.-C. (1985). *Elements of Psychophysical Theory.* New York: Oxford University Press.

Fechner, G. T. (1860). *Elemente der Psychophysik.* Leipzig: Breitkopf und Härtel.

Fechner, G. T. (1965) *Elements of Psychophysics,* Vol. 1, transl. H. E. Adler. New York: Holt, Rinehart & Winston.

Feller, W. (1967). *Probability Theory and its Applications.* Vol. 1, 3rd edn. New York: Wiley.

Ferris, G. E. (1957). A modified latin square design for taste-testing. *Food Research* **22**, 251–8.

Ferris, G. E. (1958). The *k*-visit method of consumer testing. *Biometrics* **14**, 39–49.

Fienberg, S. E. (1979). Log linear representation for paired comparison models with ties and within-pair order effects *Biometrics* **35**, 479–81.

Fienberg, S. E. and Larntz, K. (1976). Log linear representation for paired and multiple comparisons models. *Biometrika* **63**, 245–54.

Fisher, R. A. and Yates, F. (1957). *Statistical Tables for Biological, Agricultural and Medical Research.* 5th edn, London: Oliver & Boyd.

Fleckenstein, M., Freund, R. A., and Jackson, J. E. (1958). A paired comparison test of typewriter carbon papers. *Tappi* **41**, 128–30.

Flueck, J. A. and Korsh, J. F. (1974). A branch search algorithm for maximum likelihood paired comparison ranking. *Biometrika* **61**, 621–6.

Flueck, J. A. and Korsh, J. F. (1975). A generalized approach to maximum likelihood paired comparison rankings. *Ann. Statist.* **3**, 846–61.

Ford, L. R., Jr. (1957). Solution of a ranking problem from binary comparisons. *Amer. Math. Monthly* **64**, 28–33.

Ford, L. R., Jr. and Johnson, S. M. (1959). A tournament problem. *Amer. Math. Monthly* **66**, 387–9.

Fox, M. (1973). Double elimination tournaments (Letter to the Editor). *Amer. Statist.* **27**, 90–91.

Freund, J. E. (1956). Round robin mathematics. *Amer. Math. Monthly* **63**, 112–14.

Frijters, J. E. R. (1979). Variations of the triangular method and the relationship of its unidimensional probabilistic models to three-alternative forced-choice signal detection theory models. *Brit. J. Math. Statist. Psychol.* **32**, 229–41.

Frijters, J. E. R. (1982). Expanded tables for conversion of a proportion of correct responses (P_c) to the measure of sensory difference (d') for the triangular method and the 3-alternative forced choice procedure. *J. Food Science* **47**, 139–43.

Fulkerson, D. R. (1965). Upsets in round robin tournaments. *Canad. J. Math.* **17**, 957–69.

Gacula, M. C., Jr. and Singh. J. (1984). *Statistical Methods in Food and Consumer Research.* London: Academic Press.

George, S. L. (1974). Replicated paired comparisons between 2 objects – (K, R) – series. *J. Amer. Statist. Assoc.* **69**, 750–4.

Gerard, H. B. and Shapiro, H. N. (1958). Determining the degree of inconsistency in a set of paired comparisons. *Psychometrika* **23**, 33–46.

Gibbons, J. D., Olkin, I., and Sobel, M. (1977). *Selecting and Ordering Populations.* New York: Wiley.

Gilbert, E. N. (1961). Design of mixed doubles tournament. *Amer. Math. Monthly* **68**, 124–31.

Gillot, Ch. and Caussinus, H. (1966). Sur un modèle de comparaisons par paires avec une échelle de réponses à trois valeurs. *Rev. Statist. Appl.* **14**, 31–42.

Girshick, M. A., Mosteller, F., and Savage, L. J. (1946). Unbiased estimates for certain binomial sampling problems with applications. *Ann. Math. Statist.* **17**, 13–23.

Glenn, W. A. (1960). A comparison of the effectiveness of tournaments.

Biometrika **47**, 253–62.
Glenn, W. A. and David, H. A. (1960). Ties in paired-comparison experiments using a modified Thurstone–Mosteller model. *Biometrics* **16**, 86–109.
Good, I. J. (1955). On the marking of chess-players. *Math. Gaz.* **39**, 292–6.
Greenberg, M. G. (1965). A modification of Thurstone's law of comparative judgment to accommodate a judgment category of "equal" or "no difference." *Psychol. Bull.* **64**, 108–12.
Gridgeman, N. T. (1959). Pair comparison, with and without ties. *Biometrics* **15**, 382–8.
Gridgeman, N. T. (1963). Significance and adjustment in paired comparisons. *Biometrics,* **19**, 213–28.
Gridgeman, N. T. (1970). A reexamination of the two-stage triangle test for perception of sensory differences. *J. Food Science* **35**, 87–91.
Grizzle, J. E., Starmer, C. F., and Koch, G. G. (1969). Analysis of categorical data by linear models. *Biometrics* **25**, 489–504.
Gulliksen, H. (1956). A least squares solution for paired comparisons with incomplete data. *Psychometrika* **21**, 125–34.
Gulliksen, H. and Tucker, L R. (1961). Ageneral procedure for obtaining paired comparisons from multiple rank orders. *Psychometrika* **26**, 173–83.
Gumbel, E. J. (1961). Sommes et différences de valeurs extrêmes indépendantes. *C. R. Acad. Sci. Paris* **253**, 2838–9.
Gupta, S. C. (1987). Generating generalized cyclic designs with factorial balance. *Comm. Statist. – Theory Methods A* **16**, 1885–1900.
Gupta, S. S. and Sobel, M. (1960). Selecting a subset containing the best of several binomial populations. In *Contributions to Probability and Statistics,*ed. I. Olkin *et al.* Stanford: University Press, 224–48.
Guttman, L. (1946). An approach for quantifying paired comparisons and rank order. *Ann. Math. Statist.* **17**. 144–63.
Handa, B. R. and Maitri, V. (1984). On a knockout selection procedure. *Sankhyā A* **46**, 267–76.
Harary, F., Norman, R. Z., and Cartwright, D. (1965). *Structural Models: an Introduction to the Theory of Directed Graphs.* New York: Wiley.
Harris, W. P. (1957). A revised law of comparative judgment. *Psychometrika* **22**, 189–98.
Hartigan, J. A. (1966). Probabilistic completion of a knockout tournament. *Ann. Math. Statist.* **37**, 495–503.
Hartley, H. O. (1950). The use of range in analysis of variance. *Biometrika* **37**, 271–80.
Harvard University Computation Laboratory (1955). *Tables of the Cumulative Binomial Distribution.* Cambridge, Mass.: Harvard University Press.
Haselgrove, J. and Leech, J. (1977). A tournament design problem. *Amer. Math. Monthly* **84**, 198–201.
Hasse, M. (1961). Über die Behandlung graphentheoretischer Probleme unter Verwendung der Matrizenrechnung. *Wiss. Zeit. Tech. Univ.*

Dresden **10**, 1313–16.

Hemelrijk, J. (1952). A theorem on the sign test when ties are present. *Indag. Math.* **14**, 322–6.

Hoeffding, W. (1956). On the distribution of the number of successes in independent trials. *Ann. Math. Statist.* **27**, 713–21.

Hopkins, J. W. and Gridgeman N. T. (1955). Comparative sensitivity of pair and triad flavor intensity difference tests. *Biometrics* **11**, 63–8.

Horsnell, G. (1969). A theory of consumer behaviour derived from repeat paired preference testing (with Discussion). *J. R. Statist. Soc. A* **132**, 164–93.

Horsnell, G. (1977). Paired comparison product testing when individual preferences are stochastic: an alternative model. *Appl. Statist.* **26**, 162–72.

Huber, P. J. (1963a). A remark on a paper of Trawinski and David entitled: "Selection of the best treatment in a paired-comparison experiment". *Ann. Math. Statist.* **34**, 92–4.

Huber, P. J. (1963b). Pairwise comparison and ranking: optimum properties of the row sum procedure. *Ann. Math. Statist.* **34**, 511–20.

Hubert, L. J. and Golledge, R. G. (1981). Matrix reorganization and dynamic programming: applications to paired comparisons and unidimensional seriation. *Psychometrika* **46**, 429–41.

Hubert, L. and Schultz, J. (1975). Maximum likelihood paired-comparison ranking and quadratic assignment. *Biometrika* **62**, 655–9.

Hutchinson, T. P. (1979). A comment on replicated paired comparisons. *Appl. Statist.* **28**, 163–9.

Hwang, F. K. (1982). New concepts in seeding knockout tournaments. *Amer. Math. Monthly* **89**, 235–9.

Hwang, F. K. and Hsuan, F. (1980). Stronger players win more knockout tournaments in average. *Comm. Statist.-Theor. Meth. A* **9**, 107–13.

Imrey, P. B., Johnson, W. D., and Koch, G. G. (1976). Incomplete contingency table approach to paired-comparison experiments. *J. Amer. Statist. Assoc.* **71**, 614–23.

Israel, R. B. (1981). Stronger players need not win more knockout tournaments. *J. Amer. Statist. Assoc.* **76**, 950–51.

Jablonski, E. (1892). Théorie des permutations et des arrangements circulaires complets. *J. Math. Pure Appl.,* 4th series **8**, 331–49.

Jackson, J. E. and Fleckenstein, M. (1957). An evaluation of some statistical techniques used in the analysis of paired comparison data. *Biometrics* **13**, 51–64.

John, J. A. (1967). Reduced group divisible paired comparison designs. *Ann. Math. Statist.* **38**, 1887–93.

John, J. A. (1981). Efficient cyclic designs. *J. R. Statist. Soc. B* **43**, 76–80.

John, J. A. (1986). *Cyclic Designs.* London: Chapman and Hall.

John, J. A., Wolock, F. W., and David, H. A. (1972). Cyclic designs. *Nat. Bur. Stand. Appl. Math. Ser.* **62** (Washington, D. C.).

Kadane, J. B. (1966). Some equivalence classes in paired comparisons. *Ann. Math. Statist.* **37**, 488-94.

Kaiser, H. F. and Serlin, R. C. (1978). Contributing to the method of

paired comparisons. *Appl. Psychol. Meast.* **2**, 423–32.
Kalmijn, W. M. (1976). Simultaneous confidence regions for repeat preference testing. *Appl. Statist.* **25**, 117–22.
Katz, L. (1977). An efficient sequential ranking procedure. *J. Amer. Statist. Assoc.* **72**, 841–4.
Kempthorne, O. (1953). A class of experimental designs using blocks of two plots. *Ann. Math. Statist.* **24**, 76–84.
Kendall, M. G. (1955). Further contributions to the theory of paired comparisons. *Biometrics* **11**, 43–62.
Kendall, M. G. (1962). *Rank correlation methods*, 3rd edn. London: Griffin.
Kendall, M. G. and Smith, B. Babington (1940). On the method of paired comparisons. *Biometrika* **31**, 324–45.
Kendall, M. G. and Stuart, A (1977). *The Advanced Theory of Statistics*, Vol. 1, 4th edn. London: Griffin; New York: Macmillan.
Knuth, D. E. (1975). *The Art of Computer Programming*. Vol. 3 (2nd printing). Reading, Massachusetts: Addison-Wesley.
Koehler, K. J. and Ridpath, H. (1982). An application of a biased version of the Bradley–Terry–Luce model to professional basketball results. *J. Math. Psychol.* **25**, 187–205.
Kousgaard, N. (1976). Models for paired comparisons with ties. *Scand. J. Statist.* **3**, 1–14.
Kousgaard, N. (1979). A conditional approach to the analysis of data from paired comparison experiments incorporating within-pair order effects. *Scand. J. Statist.* **6**, 154–60.
Kousgaard, N. (1984). Analysis of a sound field experiment by a model for paired comparisons with explanatory variables. *Scand. J. Statist.* **11**, 51–7.
Kraitchik, M. (1953). *Mathematical Recreations*. New York: Dover Publications, 230–7.
Lancaster, J. G. and Quade, D. (1983). Random effects in paired-comparison experiments using the Bradley–Terry model. *Biometrics* **39**, 245–9.
Landau, H. G. (1951). On dominance relations and the structure of animal societies: I. Effect of inherent characteristics. *Bull. Math. Biophysics* **13**, 1–19.
Landau, H. G. (1953). On dominance relations and the structure of animal societies: III. The condition for a score structure. *Bull. Math. Biophysics* **15**, 143–8.
Latta, R. B. (1979). Composition rules for probabilities from paired comparisons. *Ann. Statist.* **7**, 349–71.
Leonard, T. (1977). An alternative Bayesian approach to the Bradley–Terry model for paired comparisons. *Biometrics* **33**, 121–32.
Lewis, S. M. and Tuck, M. G. (1985). Paired comparison designs for factorial experiments *Appl. Statist.* **34**, 227–34.
Linhart, H. (1966). Streuungszerlegung für Paar-Vergleiche. *Metrika* **10**, 16–38.
Littell, R. C. and Boyett, J. M. (1977). Designs for $r \times c$ factorial paired

comparison experiments. *Biometrika* **64**, 73–7.

Loflin, G. and McMahan, G. A. (1982). A computer program for maximum likelihood paired comparison rankings. *Computer Programs in Biomedicine* **14**, 47–52.

Luce, R. D. (1959). *Individual Choice Behavior.* New York: Wiley.

Luce R. D. (1961). A choice theory analysis of similarity judgments. *Psychometrika* **26**, 151–63.

Luce, R. D. (1977). The choice axiom after twenty years. *J. Math. Psychol.* **15**, 215–33.

Mallows, C. L. (1957). Non-null ranking models. I. *Biometrika* **44**, 114–30.

Manacher, G. K. (1979). The Ford–Johnson sorting algorithm is not optimal. *J. ACM* **26**, 441–56.

Mattenklott, A., Sehr, J., and Miescke, K.-J. (1982). A stochastic model for paired comparisons of social stimuli. *J. Math. Psychol.* **26**, 149–68.

Maurer, W. (1975). On most effective tournament plans with fewer games than competitors. *Ann. Statist.* **3**, 717–27.

McCormick, E. J. and Bachus, J. A. (1952). Paired comparison ratings: 1. The effect on ratings of reductions in the number of pairs. *J. Appl. Psychol.* **36**, 123–7.

McCormick, E. J. and Roberts, W. K. (1952). Paired comparison ratings: 2. The reliability of ratings based on partial pairings. *J. Appl. Psychol.* **36**, 188–92.

McKeon, J. J. (1960). Some cyclic incomplete paired comparisons designs. Tech. Rep. No. 24, Psychom. Lab., Univ. of North Carolina.

Miller, R. G. (1981). *Simultaneous Statistical Inference,* 2nd edn. New York: Springer.

Mohanty, S. G. (1979). *Lattice Path Counting and Applications.* London: Academic Press.

Moon, J. W. (1968). *Topics on Tournaments.* New York: Holt, Rinehart and Winston.

Moon, J. W. (1970). The expected strength of losers in knockout tournaments. In *Combinatorial Structures and their Applications,* ed. R. Guy *et al.* New York: Gordon and Breach, 277–82.

Moon, J. W. and Pullman, N. J. (1970). On generalized tournament matrices *SIAM Rev.* **12**, 384–99.

Moran, P. A. P. (1947). On the method of paired comparisons. *Biometrika* **34**, 363–5.

Morrison, H. W. (1963). Testable conditions for triads of paired comparison choices. *Psychometrika* **28**, 369–90.

Mosteller, F. (1951a, b, c). Remarks on the method of paired comparisons: I. The least squares solution assuming equal standard deviations and equal correlations. II. The effect of an aberrant standard deviation when equal standard deviations and equal correlations are assumed. III. A test of significance for paired comparisons when equal standard deviations and equal correlations are assumed. *Psychometrika* **16**, 3–9, 203–6, 207–18.

Mosteller, F. (1952). The world series competition. *J. Amer. Statist.*

Assoc. **47**, 355–400.
Mosteller, F. (1958). The mystery of the missing corpus. *Psychometrika* **23**, 279–89.
Narayana, T. V. (1968) Quelques résultats relatifs aux tournois "knockout" et leurs applications aux comparaisons aux paires. *C. R. Acad. Sci. Paris* **267**, 32–3.
Narayana, T. V. (1979). *Lattice Path Combinatorics with Statistical Applications.* Toronto: University Toronto Press.
Narayana, T. V. and Agyepong. (1980). Contributions to the theory of tournaments IV. A comparison of tournaments through probabilistic completion. *Cahiers BURO, Paris.*
Narayana, T. V. and Bent, D. H. (1964). Computation of the number of score sequences in round-robin tournaments. *Canad. Math. Bull.* **7**, 133–6.
Narayana, T. V. and Hill, J. (1974). Contributions to the theory of tournaments III. *Proc. 5th Nat. Math. Conf.,* Shiraz, Iran, 187–221.
Nishisato, S. (1978). Optimal scaling of paired comparison and rank order data: an alternative to Guttman's formulation. *Psychometrika* **43**, 263–72.
Nishisato, S. (1980). *Analysis of Categorical Data: Dual Scaling and its Applications.* Toronto: Univ. Toronto Press.
Noether, G. E. (1960). Remarks about a paired comparison model. *Psychometrika* **25**, 357–67.
Paterson, L. J. (1983). Circuits and efficiency in incomplete block designs. *Biometrika* **70**, 215–25.
Paterson, L. J. and Wild, P. (1986). Triangles and efficiency factors. *Biometrika* **73**, 289–99.
Pearson, E. S. and Hartley, H. O. (1951). Charts of the power function for analysis of variance tests, derived from the non-central *F*-distribution. *Biometrika* **38**, 112–30; also Table 30 in Pearson and Hartley (1972).
Pearson, E. S. and Hartley, H. O. (1970). *Biometrika Tables for Statisticians.* Vol. I, 3rd edn. (with additions). Cambridge: University Press; London, Griffin.
Pearson, E. S. and Hartley, H. O. (1972, 2nd imp. 1976). *Biometrika Tables for Statisticians.* Vol. II. Cambridge: University Press; London, Griffin.
Pendergrass, R. N. and Bradley, R. A. (1960). Ranking in triple comparisons. *Contributions to Probability and Statistics,* ed. I. Olkin *et al.* Stanford: University Press, 331–51.
Pfanzagl, J. (1971). *Theory of Measurement.* 2nd edn. Würzburg–Wien: Physica-Verlag.
Phillips, J. P. N. (1967). A procedure for determining Slater's *i* and all nearest adjoining orders. *Brit. J. Math. Statist. Psychol.* **20**, 217–25.
Philips, J. P. N. (1969). A further procedure for determining Slater's *i* and all nearest adjoining orders. *Brit. J. Math. Statist. Psychol.* **22**, 97–101.
Putter, J. (1955). The treatment of ties in some nonparametric tests. *Ann. Math. Statist.* **26**, 368–86.
Quenouille, M. H. and John, J. A. (1971). Paired comparison designs for

2^n factorials. *Appl. Statist.* **20**, 16–24.
Ramanujacharyulu, C. (1964). Analysis of preferential experiments. *Psychometrika* **29**, 257–61.
Ranyard, R. H. (1976). An algorithm for maximum likelihood ranking and Slater's *i* from paired comparisons. *Brit. J. Math. Statist. Psychol.* **29**, 242–8.
Rao, P. V. and Kupper, L. L. (1967). Ties in paired-comparison experiments: a generalization of the Bradley–Terry model. *J. Amer. Statist. Assoc.* **62**, 194–204.
Rapoport, A. (1949). Outline of a probabilitic approach to animal sociology: I. *Bull. Math. Biophysics* **11**, 183–96.
Remage, R., Jr. and Thompson, W. A., Jr. (1966). Maximum-likelihood paired comparison rankings. *Biometrika* **53**, 143–9.
Ross, R. T. (1934). Optimum orders for the presentation of pairs in the method of paired comparisons. *J. Educ. Psychol.* **25**, 375–82.
Rubinstein, A. (1980). Ranking the participants in a tournament. *SIAM J. Appl. Math.* **38**, 108–11.
Russell, K. G. (1980). Balancing carry-over effects in round robin tournaments. *Biometrika* **67**, 127–32.
Ryser, H. J. (1964). Matrices of zeros and ones in combinatorial mathematics. In *Recent Advances in Matrix Theory,* ed. H. Schneider. Madison: University of Wisconsin Press, 103–24.
Sadasivan, G. (1983). Within-pair order effects in paired comparisons. *Studia Sci. Math. Hungar.* **18**, 229–38.
Scheffé, H. (1952). An analysis of variance for paired comparisons. *J. Amer. Statist. Assoc.* **47**, 381–400.
Scheffé, H. (1953). A method for judging all contrasts in the analysis of variance. *Biometrika* **40**, 87–104.
Scheid, F. (1960). A tournament problem. *Amer. Math. Monthly* **67**, 39–41.
Searle, S. R. (1982). *Matrix Algebra Useful for Statistics.* New York: Wiley.
Searls, D. T. (1963). On the probability of winning with different tournament procedures. *J. Amer. Statist. Assoc.* **58**, 1064–81.
Sen, P. K. and David, H. A. (1968). Paired comparisons for paired characteristics. *Ann. Math. Statist.* **39**, 200–8.
Seneta, E. (1973). *Non-Negative Matrices.* New York: Wiley.
Simmons, G. J. and Davis, J. A. (1975). Pair designs. *Comm. Statist.* **4**, 255–72.
Singh, J. (1976). A note on paired comparison rankings. *Ann. Statist.* **4**, 651–4.
Singh, J. and Gupta, R. S. (1975). Derivation of a paired comparison model. In *Applied Statistics,* ed. R. P. Gupta. Amsterdam: North-Holland, 295–300.
Singh, J. and Gupta, R. S. (1978). A paired comparison model allowing for ties. *Scand. J. Statist.* **5**, 65–8.
Singh, J. and Thompson, W. A., Jr. (1968). A treatment of ties in paired comparisons. *Ann. Math. Statist.* **39**, 2002–15.
Sirotnik, B. W. and Beaver, R. J. (1984). Paired comparison experiments

involving all possible pairs. *Brit. J. Math. Statist. Psychol.* **37**, 22–33.
Slater, P. (1961). Inconsistencies in a schedule of paired comparisons. *Biometrika* **48**, 303–12.
Sobel, M. and Weiss, G. H. (1970). Inverse sampling and other selection procedures for tournaments with 2 or 3 players. *Nonparametric Techniques in Statistical Inference*, ed. M. L. Puri. Cambridge: University Press, 515–43.
Spence, I. and Domoney, D. W. (1974). Single subject incomplete designs for nonmetric multidimensional scaling. *Psychometrika* **39**, 469–90.
Springall, A. (1973). Response surface fitting using a generalization of the Bradley–Terry paired comparison model. *Appl. Statist.* **22**, 59–68.
Starks, T. H. (1958). Significance tests in experiments involving paired comparisons. Unpublished Ph.D. dissertation, Virginia Poly. Inst.
Starks, T. H. (1982). A Monte Carlo evaluation of response surface analysis based on paired-comparison data. *Comm. Statist. B* **11**, 603–17.
Starks, T. H. and David, H. A. (1961). Significance tests for paired-comparison experiments. *Biometrika* **48**, 95–108.
Steinhaus, H. (1950). *Mathematical Snapshots*. Oxford: University Press.
Steinhaus, H. and Trybula, S. (1959). On a paradox in applied probabilities. *Bull. Acad. Polon. Sci.* série sci. math. astr. phys. **7**, 67–9.
Stigler, S. M. (1986). *The History of Statistics: the Measurement of Uncertainty before 1900.* Cambridge, Mass.: Harvard University Press.
Thompson, G. L. (1958). Lectures on game theory, Markov chains and related topics. *Sandia Corporation Monograph* SCR – 11.
Thompson, W. A., Jr. and Remage. R., Jr. (1964). Rankings from paired comparisons. *Ann. Math. Statist.* **35**, 739–47.
Thompson, W. A., Jr. and Singh, J. (1967). The use of limit theorems in paired comparison model building. *Psychometrika* **32**, 255–64.
Thurstone, L. L. (1927a). A law of comparative judgment. *Psychol. Rev.* **34**, 273–86.
Thurstone, L. L. (1927b). Psychophysical analysis. *Amer. J. Psychol.* **38**, 368–89.
Thurstone, L. L. (1927c). The method of paired comparisons for social values. *J. Abnorm. Soc. Psychol.* **21**, 384–400.
Torgerson, W. S. (1958). *Theory and Methods of Scaling.* New York: Wiley.
Trawinski, B. J. (1961). Frequencies of partitions for paired-comparison experiments and related tables. Virginia Poly. Inst., Tech. Rep. 52.
Trawinski, B. J. (1985). Expected size of selected subset in paired comparison experiments. In *Selected Tables in Mathematical Statistics* **8**, 1–39.
Trawinski, B. J. and David, H. A. (1963). Selection of the best treatment in a paired-comparison experiment. *Ann. Math. Statist.* **34**, 75–91.
Trybula, S. (1961). On the paradox of three random variables. *Zastosowania Mat.* **5**, 321–32.
Tutz, G. (1986). Bradley–Terry–Luce models with an ordered response. *J. Math. Psychol.* **30**, 306–16.

Tversky, A. (1969). Intransitivity of preferences. *Psychol. Rev.* **76**, 31–8.

Ura, S. (1957). On Scheffé's analysis of variance for paired comparisons. *Rep. Stat. Appl. Res., JUSE* **4**, 132–46.

Ura, S. (1960a). Pair, triangle and duo-trio test. *Rep. Stat. Appl. Res., JUSE* **7**, 107–19.

Ura, S. (1960b). Selection of judges by ranking method. *Rep. Stat. Appl. Res., JUSE* **7**, 120–30.

Usiskin, Z. (1964). Max-min probabilities in the voting paradox. *Ann. Math. Statist.* **35**, 857–62.

van Baaren, A. (1978). On a class of extensions to the Bradley–Terry model in paired comparisons. *Statist. Neerlandica* **32**, 57–66.

van der Heiden, J. A. (1952). On a correction term in the method of paired comparisons. *Biometrika* **39**, 211–12.

van Elteren, P. and Noether, G. E. (1959). The asymptotic efficiency of the χ_r^2-test for a balanced incomplete block design. *Biometrika* **46**, 475–7.

van Putten, W. L. J. (1982). Maximum likelihood estimation for Luce's choice model. *J. Math. Psychol.* **25**, 163–74.

Wei, T. H. (1952). The algebraic foundations of ranking theory. Unpublished thesis, Cambridge University.

Wierenga, B. (1974). Paired comparison product testing when individual preferences are stochastic. *Appl. Statist.* **23**, 384–96.

Wilkinson, J. W. (1957). An analysis of paired comparison designs with incomplete repetitions. *Biometrika* **44**, 97–113.

Williams, E. R. (1976). Resolvable paired-comparison designs. *J. R. Statist. Soc. B* **38**, 171–4.

Winsberg, S. and Ramsay, J. O. (1981). Analysis of pairwise preference data using integrated *B*-splines. *Psychometrika* **46**, 171–86.

Yalavigi, C. C. (1967). A mixed doubles tournament problem. *Amer. Math. Monthly* **74**, 926–33. Correction **75**, 623.

Yang, S. S. (1977). General distribution theory of the concomitants of order statistics. *Ann. Statist.* **5**, 996–1002.

Yellott, J. I., Jr. (1977). The relationship between Luce's choice axiom, Thurstone's theory of comparative judgment, and the double exponential distribution. *J. Math. Psychol.* **15**, 109–44.

Zermelo, E. (1929). Die Berechnung der Turnier-Ergebnisse als ein Maximumproblem der Wahrscheinlichkeitsrechnung. *Math. Zeit.* **29**, 436–60.

Zimmermann, H. and Rahlfs, V. W. (1976). Die Erweiterung des Bradley–Terry–Luce modells auf Bindungen im Rahmen verallgemeinerter (Log)-Linearmodelle. *Biom. Zeit.* **18**, 23–32.

INDEX